Purinergic System

Edited by Margarete Dulce Bagatini

Published in London, United Kingdom

Purinergic System
http://dx.doi.org/10.5772/intechopen.94827
Edited by Margarete Dulce Bagatini

Contributors
Yerly Magnolia Useche Salvador, Lucianne Fragel Madeira, Douglas Penaforte Cruz, Marinna Garcia Repossi, Richa Rai, Juan Pedro Lapuente, Shin Mukai, Arinzechukwu Ude, Kelechi Okeke, Hailian Shen, Margarete Dulce Bagatini, Daiane Manica, Marcelo Moreno, Gilnei Bruno da Silva

First published in London, United Kingdom, 2022 by IntechOpen
IntechOpen is the global imprint of INTECHOPEN LIMITED, registered in England and Wales, registration number: 11086078, 5 Princes Gate Court, London, SW7 2QJ, United Kingdom

British Library Cataloguing-in-Publication Data
A catalogue record for this book is available from the British Library

Additional hard and PDF copies can be obtained from orders@intechopen.com

Purinergic System
Edited by Margarete Dulce Bagatini
p. cm.

This title is part of the Biochemistry Book Series, Volume 36
Topic: Metabolism
Series Editor: Miroslav Blumenberg
Topic Editor: Yannis Karamanos

Print ISBN 978-1-80355-783-0
Online ISBN 978-1-80355-784-7
eBook (PDF) ISBN 978-1-80355-785-4
ISSN 2632-0983

IntechOpen Book Series
Biochemistry
Volume 36

Aims and Scope of the Series

Biochemistry, the study of chemical transformations occurring within living organisms, impacts all of the life sciences, from molecular crystallography and genetics, to ecology, medicine and population biology. Biochemistry studies macromolecules - proteins, nucleic acids, carbohydrates and lipids –their building blocks, structures, functions and interactions. Much of biochemistry is devoted to enzymes, proteins that catalyze chemical reactions, enzyme structures, mechanisms of action and their roles within cells. Biochemistry also studies small signaling molecules, coenzymes, inhibitors, vitamins and hormones, which play roles in the life process. Biochemical experimentation, besides coopting the methods of classical chemistry, e.g., chromatography, adopted new techniques, e.g., X-ray diffraction, electron microscopy, NMR, radioisotopes, and developed sophisticated microbial genetic tools, e.g., auxotroph mutants and their revertants, fermentation, etc. More recently, biochemistry embraced the 'big data' omics systems. Initial biochemical studies have been exclusively analytic: dissecting, purifying and examining individual components of a biological system; in exemplary words of Efraim Racker, (1913 –1991) "Don't waste clean thinking on dirty enzymes." Today, however, biochemistry is becoming more agglomerative and comprehensive, setting out to integrate and describe fully a particular biological system. The 'big data' metabolomics can define the complement of small molecules, e.g., in a soil or biofilm sample; proteomics can distinguish all the proteins comprising e.g., serum; metagenomics can identify all the genes in a complex environment e.g., the bovine rumen.
This Biochemistry Series will address both the current research on biomolecules, and the emerging trends with great promise.

Meet the Series Editor

Miroslav Blumenberg, Ph.D., was born in Subotica and received his BSc in Belgrade, Yugoslavia. He completed his Ph.D. at MIT in Organic Chemistry; he followed up his Ph.D. with two postdoctoral study periods at Stanford University. Since 1983, he has been a faculty member of the RO Perelman Department of Dermatology, NYU School of Medicine, where he is codirector of a training grant in cutaneous biology. Dr. Blumenberg's research is focused on the epidermis, expression of keratin genes, transcription profiling, keratinocyte differentiation, inflammatory diseases and cancers, and most recently the effects of the microbiome on the skin. He has published more than 100 peer-reviewed research articles and graduated numerous Ph.D. and postdoctoral students.

Meet the Volume Editor

Dr. Margarete Dulce Bagatini is an associate professor at the Federal University of Fronteira Sul (UFFS), Brazil. She has a degree in Pharmacy and a Ph.D. in Biological Sciences (Toxicological Biochemistry). She is a member of the UFFS research advisory committee and the Biovitta Research Institute. She is currently the leader of the Biological and Clinical Studies in Human Pathologies research group. She is also a professor of postgraduate programs in Biomedical Sciences and Science and Food Technology at UFFS and the postgraduate program in Biochemistry at the Federal University of Santa Catarina (UFSC), Brazil. She has experience in pharmacy and clinical analysis, including in oxidative stress, the purinergic system, and human pathologies. Dr. Bagatini is a reviewer of several international journals and books.

Contents

Preface

The purinergic system is a set of components capable of generating a cellular intercommunication network. Composed of signaling molecules, regulatory enzymes, and specific receptors, this organization can modulate several basal pathways of the organism. However, its understanding is relatively recent for researchers. In 1929, Drury and Szent-Gyorgyi described the extracellular functions associated with purines. In 1972, Burnstock defined the concept of purinergic neurotransmission, given that nucleotides and nucleosides were studied only in energy activities. In 1978, the extracellular receptors of the P1 and P2 groups were described in the literature by Burnstock when empirically demonstrating the role of adenosine triphosphate (ATP) as a transmitter and other functions in some processes.

Characterized as a common signaling pathway between cells, the purinergic system is capable of modulating physiological and biochemical processes. It has been widely studied for its therapeutic potential and modulation of physiological processes, such as performance in the process of apoptosis, thromboregulation, cell proliferation, platelet aggregation, endothelial vasodilation, and pain mechanism as well as neurotransmission and neuroprotection. In addition, several studies report an association of the purinergic system with inflammatory processes and immune responses.

In an initial approach, it is necessary to define the executing components of this system, given that its signaling is dependent on the biochemical interactions of these structures and conducted by purines. As for neurotransmitters, ATP, adenosine diphosphate (ADP), and adenosine monophosphate (AMP) constitute the nucleotides, while there is the adenosine nucleoside. Specific enzymes regulate these ectoenzymes, divided into adenosine deaminase (ADA) responsible for deaminating adenosine to inosine and ectonucleotidases E-NPP, E-NTPDases, ecto-5'-nucleotidase, and alkaline phosphates that hold nucleotides.

Nucleosides and nucleotides are signaling biomolecules that have functional diversity in the extracellular environment and modulate several biological effects. Its differences are noticeable initially by the number of phosphate groups and may have three groups (ATP) or be absent (adenosine). However, its regulation and reception also change. In this view, the receivers are from different organizations and initially differ in groups P1 and P2. The first is closely related to the nucleoside adenosine and are channels coupled to the G protein subdivided into A1, A2a, A2b, and A3. The second group has a preference for mono-, di-, and triphosphate nucleotides, such as AMP, ADP, and ATP, being subdivided into P2X, linked to ion channels that, when activated, result in the opening of pores in the cell membrane and the passage of Na+, K+, and cations Ca2+.

All cells have components of the purinergic system and could release nucleotides in a controlled manner. The mechanisms related to the release of nucleotides have been an intense focus of research activities. Studies have extensively investigated the physiology, pharmacology, and biochemistry of purinergic signaling. More recently, the focus has been on the pathophysiology and therapeutic potential of components of the purinergic system, especially concerning P1 and P2 receptors. It is currently established that changes in purinergic signaling are involved in the pathophysiology and therapy of many diseases, possibly in addition to being related to the effects of nutraceutical molecules.

This book describes the purinergic system and the correlation between the system and the health and disease process. It is divided into three sections and eight chapters describing purinergic and immune signaling, the purinergic system and diseases, and future perspectives.

We hope that academic researchers and students can learn and share the knowledge presented here. This book is the result of several collaborating parties. We gratefully acknowledge all the authors and reviewers for their contributions and the valuable assistance of Ms. Dolores Kuzelj and Mia Vulovic at IntechOpen for their support throughout the publication process.

Margarete Dulce Bagatini
Federal University of Fronteira Sul,
Chapecó, SC, Brazil

Section 1

Purinergic and Immune Signaling

Chapter 1

Purinergic System in Immune Response

Yerly Magnolia Useche Salvador

Abstract

In mammalian cells, the purinergic signaling and inflammatory mediators regulate each other. During microbial infection, nucleotides and nucleosides from both dying host cells and pathogens may be recognized by the host receptors. These receptors include purinergic receptors such P2X, P2Y, and A2A, as well Toll-like receptors, and NOD-like receptors. The interaction with most of these receptors activates immune responses, including inflammasome activation, releasing of pro-inflammatory cytokines, reactive nitrogen and oxygen species production, apoptosis induction, and regulation of T cell responses. Conversely, activation of adenosine receptors is associated with anti-inflammatory responses. The magnitude of resultant responses may contribute not only to the host defense but also to the homeostatic clearance of pathogens, or even to the severe progression of infectious diseases. In this chapter, we discuss how the purinergic signaling activation upregulates or downregulates mechanisms in infectious diseases caused by the bacterial, parasite, and viral pathogens, including SARS-CoV-2. As a concluding remark, purinergic signaling can modulate not only infectious diseases but also cancer, metabolic, and cardiovascular diseases, constituting a strategy for the development of treatments.

Keywords: purinergic receptors, immune responses, inflammation, infectious disease

1. Introduction

The purinergic signaling modulates pathways of both neural and non-neural physiological processes, including immune responses, inflammation, pain, platelet aggregation, endothelium-mediated vasodilation, proliferation, and cell death [1]. Three main components are part of the system. Purinergic: extracellular nucleotides and nucleosides, their receptors, and the ectoenzymes responsible for regulating the levels of these molecules [2]. In addition, nucleotides, nucleosides, and uric acid resulting from the death of infected or injured cells are also recognized by other receptors better known for their role in pathogen recognition as Toll-like receptors (TLRs), and NOD-like receptors (NLRs) [3]. Other immune innate receptors are able to detect nucleic acids (RNA or DNA) from either phagocyted or circulating microbes, including retinoic acid-inducible gene I (RIG-I)-like receptors (RLRs), the C-type lectin receptors (CLRs), and cytosolic sensors [4,5]. All immune cells are able to recognize nucleotides as a danger signal throughout either purinergic or no purinergic receptors. Immediately various immune responses can be activated,

IntechOpen

such as pro-inflammatory cytokines secretion by macrophages, quimocine production by eosinophils, maturation of dendritic cells (DCs), as well as T and B cells costimulation [6].

Extracellular nucleotides and nucleosides are released, along with many other molecules, from dead cells. Apoptosis and necrosis are the cell death mechanisms that

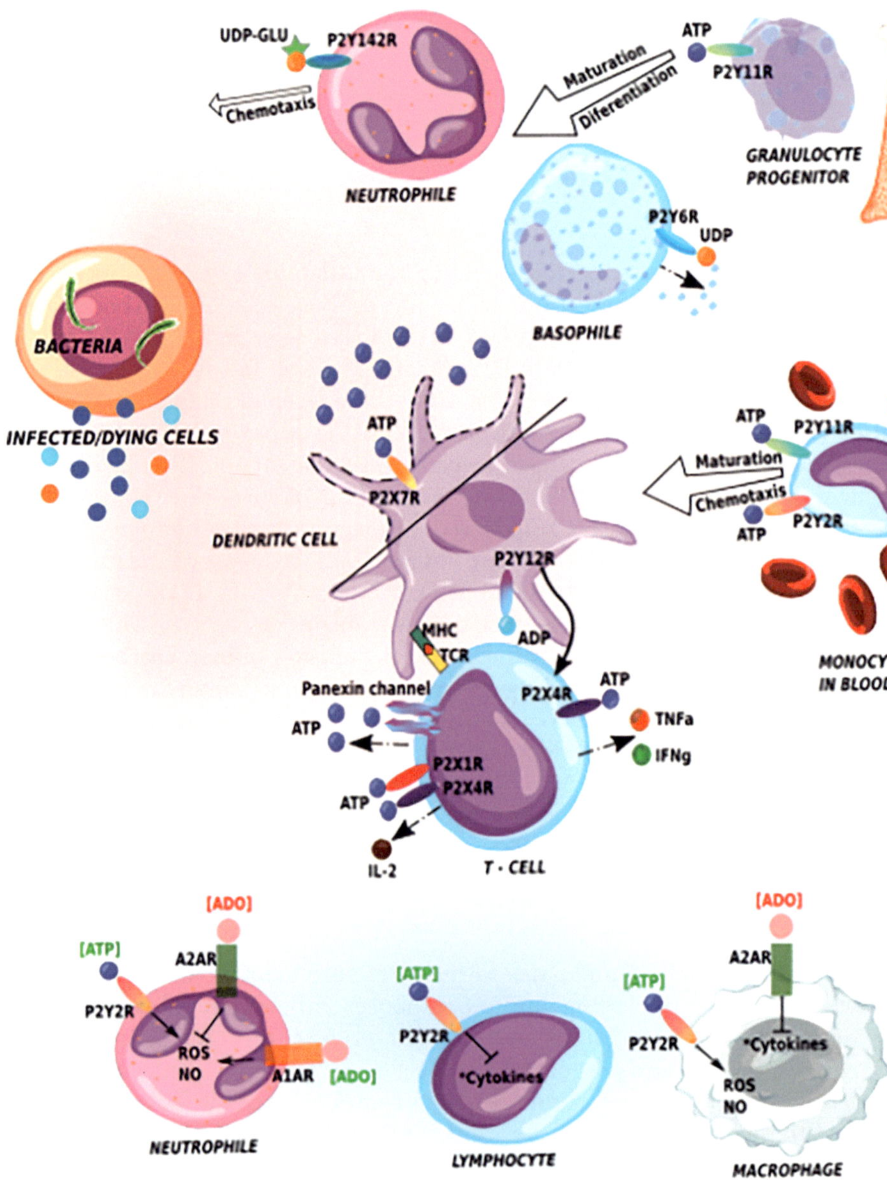

Figure 1.
*Infected cells release nucleotides and nucleosides. High ATP concentration can induce both dead cell (dashed line, in dendritic cell) and several functions over immune cells as maturation, and quimiotaxis. Low ATP concentration ([ATP]) can induce (arrow) or inhibit (bar-headed line) responses depending on the immune cell. Low adenosine concentration ([ADO]) can induce ROS and NO. High adenosine concentration ([ADO]) inhibit pro-inflammatory cytokine (*Cytokines) expression. Reddish and bluish background means pro-inflammatory and anti-inflammatory context, respectively. Source: The figure 1 design is original and cell vectors were modified from freepik's: https://www.freepik.es/vectores/personas, People Vector created by brgfx - http://www.freepik.es.*

can operate in physiological conditions. Apoptosis is activated by genetically controlled cell signals to modulate cell growth and development, as it is a programmed event. It is an ordered process that does not trigger inflammation. Conversely, necrosis is a not regulated cell death, characterized by the cell content release as a consequence of the effect of diverse environmental factors, leading to higher inflammation around [7]. Furthermore, during infections, some intracellular pathogens require cell lysis, while others have developed mechanisms to prevent cell death during their replication and dissemination outside the infected cell.

ATP is known as a damage signal, released or leaked by injured cells, or a molecular pattern associated with damage [8]. Necrotic cells may use either pannexin channels or connexin hemichannels to release intracellular ATP, and the P2X7R may be involved in this process [6]. Adenosine is a nucleoside that mediates anti-inflammatory and immunosuppressive actions, such as inhibiting the production of pro-inflammatory cytokines and lymphocyte proliferation [9]. In pathological conditions, adenosine plays a protective role acting as an endogenous regulator of innate immunity and in host defense against excessive tissue damage associated with inflammation [10]. The A1 and A2A receptors (A1R and A2AR) are activated by adenosine concentrations in the nanomolar range, while the A2B and A3 receptors (A2BR and A3R) become active only when the extracellular levels of adenosine rise in the micromolar range during periods of inflammation, hypoxia or ischemia [11] (**Figure 1**). Other nucleosides are recognized by several P2 receptors, which are going to explain later.

2. Regulation of extracellular nucleotides and nucleosides

In the context of the immune response, as mentioned, while the extracellular ATP (eATP) exhibits pro-inflammatory and stimulatory effects in the immune system, either appropriate or exacerbated responses [6], adenosine has primarily anti-inflammatory and inhibitory effects [9]. Therefore, the balance between ATP versus adenosine levels is important in modulating cellular immune responses and pathogen survival [12]. The concentrations of extracellular nucleotides and nucleosides are regulated by immune and non-immune cells through the action of enzymes anchored to the cell membrane, with their catalytic site facing the extracellular environment [13]. These enzymes, called ectonucleotidases, hydrolyze extracellular nucleotides into their respective nucleosides to control exacerbated levels of nucleotides and maintain steady-state conditions [12]. Among them, ecto-nucleoside triphosphate diphosphohydrolases (E-NTPDases, CD39, or apyrases), hydrolyzes both ATP and ADP to AMP, in the presence of divalent cations such as calcium and magnesium. Sequentially, the E-5′-nucleotidase (CD73) terminates the ectonucleotidase cascade with the hydrolysis of monophosphate nucleotides, resulting in adenosine. This in turn is hydrolyzed by the enzyme adenosine deaminase (E-ADA), transforming adenosine into inosine, its inactive metabolite [13]. In addition, the ecto-nucleotide pyrophosphatase/phosphodiesterases (E-NPPs) yield free nucleosides. The nitrogenous bases are hydrolyzed from nucleosides by the action of phosphorylases that yield ribose-1-P and free bases. If the nucleosides and/or bases are not re-utilized, the purine bases are further degraded to uric acid [3].

Two important environments for the highly reported purinergic signaling activation are the central nervous system (CNS) and the blood. The principal cell type in the brain involved in ATP degradation is the microglia through the expression of

CD39 [14]. The ATP-regulation in the blood is mediated by the red blood cells, which may regulate tissue circulation and O_2 delivery by releasing the vasodilator ATP in response to hypoxia. When released extracellularly, ATP is rapidly degraded to ADP in the circulation by ectonucleotidases. Moreover, ADP acting on P2Y13 receptors on red blood cells serves as a negative feedback pathway for the inhibition of ATP release [15].

Nucleosides and nucleotides are recognized by purinergic receptors P1, P2, TLRs, and NLRs. The recognition implies that nucleosides and nucleotides are temporarily held at different concentrations to activate their respective receptors. Purinergic receptors are divided into two families: P1 and P2 receptors [1]. The G-protein coupled metabotropic P1 receptors recognize exclusively extracellular adenosine and can be subdivided into A1R, A2AR, A2BR, and A3R. The P2 receptors can be subdivided into two subtypes: non-selective ion-gated channel P2X receptors (that recognize ATP) and G-coupled P2Y receptors (that recognize ATP, ADP, UTP, UDP, and UDP-glucose) [6]. Actually, it has been described seven P2X receptors (from P2X1 to P2X7) with different affinities for ATP. The P2X7 receptor has a low affinity for ATP (requiring $\geq$100 mM to be activated; while others can be activated at lower concentrations) [12].

The consensus about the relationship between purinergic signaling and the immune system can be summed up by the opposing effects of ATP and adenosine. The ATP contributes to triggering the inflammatory response along with molecular patterns associated with pathogens [5]. Conversely, the adenosine nucleoside mediates anti-inflammatory and immunosuppressive actions, such as inhibiting the production of pro-inflammatory cytokines and lymphocyte proliferation [9]. However, there are other nucleotides and nucleosides modulating the immune system.

3. Effects of purinergic signaling on the innate and adaptive immune system

The eATP released from stressed, dying or infected cells bind to P2 receptors (as P2X7R) and may lead to pathogen elimination through several mechanisms: (1) host cell death; (2) inflammasome activation and IL-1β secretion; and (3) production of reactive oxygen species (ROS) and nitric oxide (NO); promoting lysosome and phagosome fusion [16].

P2X7R activation is associated with pore formation, which depends on the concentration and duration of ATP treatment [17], as well as leads to the opening of pores that allow the passage of small molecules (< 900 Da) [18] as dinucleotides or nucleosides, increasing the extracellular concentration of these purinergic ligands. Inflammasomes are multi-protein complexes assembled in the host cell in response to infection or cellular stress, leading to non-homeostatic and lytic cell death, called pyroptosis. P2X7R was shown to activate NLRP3 inflammasomes [19]; and recently, caspase-11-induced pyroptosis was shown to require pannexin-1 channels and the P2X7R activation [20]. Pyroptosis is important because of the cytokines, chemokines, and damage-associated molecular patterns (DAMPs) which are released to the extracellular compartment, and also because intracellular pathogens are exposed to extracellular immune response, thus allowing their destruction [21]. At the same time, the inflammasomes lead to the maturation and secretion of pro-inflammatory cytokines, such as IL-1b and IL-18 [21]. IL-1β affects virtually all cells and organs of the body and is one of the most important cytokines that mediate autoimmunity, infections, and degenerative diseases [22]. This cytokine has a role in the CNS as an endogenous pyrogenic agent, and it can also induce inflammation, leukocyte recruitment, and Th17 profile immune responses [23]. In addition to eATP, uridin diphosphate (UDP) is released by the cleavage of pannexin-1 channels via caspase in apoptotic cells resulting from the vesicular stomatitis virus infection. Then, the UDP-P2Y6R signaling is able to protect both cells and mice from infection through an increase in IFN-β production, in acute neurotropic infection [24].

Purinergic receptors P1 (P1R) and P2 (P2R) can be expressed simultaneously in almost all immune cells, apparently depending on their ligand concentrations in the extracellular space [25]. Therefore, eATP-P2R interactions also activate pro-inflammatory responses in immune cells [26], as we have seen in infected or dying non-immune cells.

Neutrophils, granulocytes participating in both immune systems innate and adaptative, are the first immune cells in arrived at the inflammation site, constituting the main acute inflammatory response against pathogens by both phagocytosis and the oxidative microbicidal molecules production [25]. Neutrophils are the more affected cells by the purinergic signaling, probably because they express several purinergic receptors [26]. For instance, P2Y11 receptor (P2Y11R) is responsible for the ATP-mediated differentiation and maturation of granulocytic progenitors in the bone marrow [27]. Also, the interaction between eATP-P2Y11R mediates the inhibition of neutrophil apoptosis [28] and increases the chemotactic response of neutrophils [29]. In addition to eATP, other nucleotide released by damaged cells is the uridine 5′-diphosphoglucose (UDP-glucose), which activates the P2Y14R signaling. UDP-glucose promotes chemotaxis of freshly isolated human neutrophils through P2Y14R activation [30]. Moreover, some inflammatory diseases have been related to P2Y14R activation. During pelvic inflammatory disease, in the endometria in both women and female mice, the P2Y14R and pro-inflammatory cytokines as IL-8 are up-regulated in the epithelium [31]. Then, the design of therapies to modulate mucosal immunity may be done by targeting P2Y14R [30].

Occasionally, some nucleosides are more concentrated than the ATP in the extracellular matrix. When uridine triphosphate (UTP) is more available than ATP, P2Y2 receptors (P2Y2R) may be activated by either ATP or UTP mediating several activities in the P2Y2R expressing cells. For instance, fibrotic lung disease is related to some activities mediated by P2Y2R such as the lung fibroblast's proliferation and migration, the recruitment of neutrophils, and IL-6 secretion in the lungs [32]. By the other way, when the eATP-P2X1R signaling is activated in neutrophils and platelets, activated neutrophils are recruited to the injury site and their adherence to vessel walls together with the platelets occurred, promoting both thrombosis and fibrinogenesis [33].

In addition to neutrophils, the eosinophils, and basophils, other granulocyte cells, which are activated during parasitic infections and allergies, are also regulated by the purinergic system. The accumulation of eosinophils during lung inflammation is triggered by UTP-P2Y2R interaction that induces the expression of VCAM-1, an adhesion molecule, which in turn, induces changes in endothelial cell shape for the opening of passageways through which eosinophils migrate [34]. Moreover, P2Y2R activation by ATP in eosinophils has been reported to induce chemotaxis in allergic lung inflammation [35]. In other circumstances, when UDP have more concentrated than UTP or ATP, the UDP-P2Y6R signaling induces IgE-dependent degranulation in human basophils [36].

During infections, dendritic cells (DCs) are responsible for presenting antigens to naive T cells and activating them, making a link between the innate and adaptive immune response [37]. The ATP-P2R interaction is involved in the migration and differentiation of DCs [38]. Specifically, the eATP-P2Y11R interaction modulates the maturation of human monocyte-derived dendritic cells (MoDCs) [39]. Moreover, P2Y2R activation by ATP promotes chemotaxis of MoDCs [35]. In fact, the eATP-P2Y11R interaction mediates the migration of DCs accordingly with the DC type, although all DC populations express P2Y11R. MoDCs down-regulate the P2Y11R expression, decreasing the inhibition of migration triggered by ATP. While either interleukin-3 receptor-positive plasmacytoid DCs or CD1c + peripheral blood DCs do not inhibit their migration by ATP. Then, the possibility of a meeting between DCs and antigens may be mediated by gradients of ATP formed in and around inflamed areas. Therefore, after vaccination, the migration of DCs charged with antigens to near lymph nodes may be increased with the inhibition of P2Y11R expression. This strategy could improve the time of response after vaccination [38]. In addition, P2Y12 receptor (P2Y12R) modulates murine DCs function by ADP, including induction of intracellular Ca2+ transportation, macropinocytosis, and T-cell stimulation [40]. However, the stimulation of the P2X7R with ATP can induce cell death; such as in murine spleen-derived DCs, which increase the permeabilization and the intracellular calcium, resulting in apoptosis [41].

As reviewed, the T-cell activation by DCs can be modulated by the purinergic system. Additionally, lymphocytes (mainly T and B cells) that are characterized by expressing antigen receptors, allowing the activation of anti-microbial responses [25], also express purinergic receptors which modulates lymphocyte proliferation, differentiation, and functioning. Immature T cells pass through the thymus for differentiation, where stromal epithelial cells are in charge of both the positive and negative selection processes, which in turn defines the T cell functional profile between CD8+ or CD4+ cells [42]. In the thymus, P2X7R and P2Y2R are expressed in several cells as murine thymic epithelial cells (TECs), leading to the release of calcium from intracellular stores and increasing the permeabilization membrane [43]; possibly leading to TECs apoptosis as well as reported in DCs [41], and ending in the alteration of both T-cell differentiation and their peripheral functioning. Similarly, the eATP-P2X7R signaling leads to the opening of a transmembrane cationic channel that allows K+ efflux and Na + and Ca2+ influx and promotes cytoplasmic membrane depolarization in the phagocytic cell of the thymic reticulum [18], leading to increase permeabilization and apoptosis, and impairing of T-cell precursors proliferation.

During immune synapse, naïve T-cells release ATP throughout pannexin-1 channels, then the eATP interaction with P2X1R and P2X4R receptors regulates T-cell activation, calcium entry, and IL-2 release [44]. Also, γδ T-cells, abundant at barrier sites such as the skin, gut, lung, and reproductive tract, are activated and upregulated tumor necrosis factor-alpha (TNF-α), and interferon-γ (IFN-γ) release through the eATP-P2X4R interaction [45]. Moreover, P2X4Rs participate in the migration process of CD4+ T-cells. This migration is mediated by the chemokine stromal-derived factor-1α (SDF-1α), which triggers the T-cell polarization by the accumulation of ATP producing mitochondria near ATP-releasing pannexin-1 channels and newly expressed P2X4Rs. This set of molecules promotes both the Ca2+ influx and sustained mitochondrial ATP production required for the pseudopod protrusion and T-cell migration [46]. Conversely, exogenous activation of P2Y11R with eATP blocks T-cell trafficking [47].

Specifically, the purinergic system may modulate inflammatory severe clinical conditions like sepsis. This condition has two phases, first is the hyperinflammatory phase, which may be later restricted by the immunosuppressive phase. However, this last phase characterized by high blood levels of regulatory T cells (Tregs) is strongly associated with mortality. Tregs proliferation is controlled by P2Y12R activation in both Tregs and platelets, and P2Y12R blockade restores the immunological homeostasis. Therefore, this strategy may guide pharmacological treatment for sepsis and increase patient survival [48].

Among innate immune cells acting mainly in chronic inflammatory responses are the mononuclear phagocytes, which are in circulation as monocytes or in several tissues as macrophages, or in specific tissues as microglial cells in the brain. These phagocytes may have either a pro-inflammatory or anti-inflammatory role depending on the type of cytokines around them and express several P2 receptors [25]. For instance, the major pathway of macrophage activation is the eATP-P2Y11R signaling, which leads to cytokine release [49]. Moreover, macrophages exposed to LPS, increase the P2YR and P2X7R activation mediated by eATP, modulating the IL-1β, TNF-α, and NO production [50]. Monocyte adhesion process is also regulated by the ATP-P2R interaction [51]. For example, activated P2Y12R induces both vascular smooth muscle inflammatory changes via MCP-1 upregulation and monocyte adhesion into the vascular wall, promoting atherosclerotic lesions [52].

Microglia is the cell responsible for the immune function of the nervous system in both physiological and pathological conditions [53]. Increased P2X7R expression and its ATP-mediated activation in microglia are observed after the LPS brain challenge, leading to increased immune response associated with NO and ROS, along with reduced neuronal viability. Inhibition of this purinergic response may be a neuroprotective strategy in brain inflammatory diseases [54]. In addition, the upregulation of P2Y6, P2Y12, P2Y13, and P2Y14 receptors in spinal microglia is associated with the development of neuropathic pain [55]. In an inflammatory context, ADP acting on P2Y12R induces extension of microglia processes thereby attracting this cell to the site of ATP/ADP leaking or release. Moreover, the ADP-P2Y12R activation in microglia induces intracellular calcium accumulation, which in turn causes the increase of CC chemokine ligand 3 (CCL3) expression in the peripheral injured site and also in the spinal cord, inducing neuropathic pain. Since the inhibition of CCL3-CCR5 signaling suppresses the development of neuropathic pain, treatments based on inhibition of CCL3 expression can be promising to control this kind of inflammatory disorder [56].

The P2Y6R is also upregulated in microglia when neurons are damaged, then the UDP-P2Y6R signaling facilitates microglial phagocytosis [57]. Consistently, the brain injury caused by ischemic accidents is increased by the inhibition of both P2Y6R expression and the microglia-phagocytic activity [58]. The UDP-P2Y6R signaling is also associated with neuropathic pain and is partially explained by the induction of CCL2 production through the MAP kinases-NF-kappaB pathway in microglia [59]. CCL2 is a recruitment factor of myeloid cells to the regions with injured neurons. In the spinal cord, CCL2 released from primary afferent neurons and reactive astrocytes could contribute to either the induction or maintenance of chronic pain [60], neurodegenerative and neuroinflammatory diseases [61].

As mentioned above, in addition to purinergic receptors, other receptors are able to recognize eATP such as the NLRs and TLRs. These receptors recognize both pathogen-associated molecular patterns (PAMPs) and damage-associated molecular patterns (DAMPs) (including eATP and uric acid). The relationship between these receptors and nucleotides or nucleosides, and the structure and functions of NLRs will be addressed immediately. TLRs will be explained through the case of gout disease.

The NLRs are cytoplasmic receptors and the structure has three domains: a common domain organization with a central conserved domain NOD (NACHT: NAIP, CIITA, HET-E, and TP-2), N-terminal effector domain, and C-terminal leucine-rich repeats (LRRs) [62]. The NAIP motif (neuronal apoptosis inhibitor protein) and the CIITA motif (MHC class II transcription activator) contain a distinct predicted nucleoside triphosphatase (NTPase) domain. In addition, NTPase domain in CIITA shows highly significant sequence similarity to CARD4 (pro-apoptotic protein). Therefore, the NACHT family includes both pro-apoptotic (e.g. CARD4) and anti-apoptotic (e.g. NAIP) predicted NTPases [63]. In consequence, all NLRs could modulate apoptotic process. However, either the possible apoptotic effect or the ATPase activity of the NATCH domain and the consequence on the concentration of nucleotides and derivatives in the cytoplasm must be the subject of study.

Most NLRs recognize various ligands activating inflammatory responses. These ligands come from different sources, including microbial pathogens (peptidoglycan, flagellin, viral RNA, fungal hyphae, etc.), host cells (ATPs, uric acid, etc.), and esteril activators (alum, silica, UV radiation, skin irritants, etc.). In addition, some NLRs respond to cytokines such as interferons. The activated NLRs show various functions that can be divided into four broad categories: inflammasome formation, signaling transduction, transcription activation, and autophagy [64]. Several NLRs have been used to identify inflammasomes, depending on the receptor that recognizes the PAMPs (for example, NLRP1, NLRP3, AIM2, NLRC4), while the other group of inflammasomes can be activated by cytosolic lipopolysaccharides (LPS) derived from gram-negative bacteria. The NLRP3 inflammasome can be activated by different stimuli, such as bacterial, viral, and fungal pathogens, pore-forming toxins, crystals, silica, and DAMPs (for example, eATP) [65]. The activation of the NLRP3 inflammasome requires two signals: (1) a PAMP, such as LPS, leading to transcription of NF-kB, upregulating genes encoding pro-inflammatory cytokines, chemokines, and proteins involved in the inflammasome platform; and (2) a DAMP, such as eATP, which induces inflammasome activation after ligation to the P2X7R. Once activated, these complexes promote activation of the protease caspase-1, which cleaves pro-IL-1β and pro-IL-18 into their active forms: IL-1β and IL-18 [12].

In addition to the inflammasome activation, eATP induces ROS production. ROS are highly reactive chemicals formed from O_2 (such as peroxides, superoxides, and hydroxyl radicals) [66]. For instance, respiratory epithelial cells induce mitochondrial ROS in response to influenza infection. ROS induces the expression of type III interferon, a response associated with viral infection control [67]. Moreover, *Porphyromonas gingivalis (P. gingivalis)* infection of gingival epithelial cells induces assembly of the P2X4R and P2X7R to form a pore, pannexin-1, promoting ROS production triggered by ATP-P2X7R activation. Later the ROS can activate the NLRP3 inflammasome and caspase-1, resulting in bacteria death [68].

Some TLRs recognize in addition to PAMPs from intracellular or extracellular pathogens, others molecules like ATP and uric acid. TLRs recognizing uric acid are involved in several metabolic and cardiovascular diseases, including gout, chronic renal tubular damage, autosomal dominant polycystic kidney disease, and cartilage degeneration. Among them, the accumulation of uric acid crystals (monosodium urate - MSU -) in the joints causes arthritis which is the base of the metabolic disease called Gout [64]. The uric acid also could cause inflammatory events by the activation of the NLRP3 inflammasome and its activation induces the IL-1β formation, which leads to the development of gouty arthropathy [69]. Moreover, when the level of uric acid is higher than 6.8 mg/dL, MSU crystals are formed and are recognized by TLRs. These TLRs then activate the NALP3 inflammasome. MSU also triggers neutrophil activation and further produces immune mediators, which lead to a pro-inflammatory response [70]. In the mice model of Gout, it was found that MSU crystals are recognized by TLR2 and TLR4 with the participation of the TLR adapter protein myeloid differentiation factor 88 (MyD88) in bone marrow-derived macrophages. After recognition by these TLRs, MSU induced the production of pro-inflammatory cytokines as IL-1β, TNF-α, keratinocyte-derived cytokine/growth-related oncogene alpha (KDC/GROα), and transforming growth factor beta1(TGF-β1). Moreover, neutrophil influx, local induction of IL-1β, and more pro-inflammatory reaction were promoted [71].

In addition, patients with hyperuricemia (high levels of uric acid in the blood) develop vascular diseases associated with the formation of aminocarbonyl radicals from excess uric acid with the concomitant oxidative effect [72]. Additionaly, metabolic and cardiovascular diseases are associated with the activation of the renin-angiotensin system (RAS) mediated by uric acid, in adipose tissue. For instance, high blood pressure and increased expression of both TLR2/4, pro-inflammatory cytokines (TNF-α and IL-6), and RAS activation in adipocytes were found in hyperuricemic rats. These high levels of cytokines and RAS components were reverted by TLR2/4 RNA silencing [73]. Proinflammatory pathways are induced by uric acid and angiotensin II-mediated by TLR4 in renal proximal tubular cells developing chronic tubular damage [74]. Moreover, TLR-2 and TLR-4 gene expressions are associated with rapid progression in autosomal dominant polycystic kidney disease (ADPKD) patients [75]. Furthermore, in human chondrocytes, the accumulation of both calcium pyrophosphate dihydrate (CPPD) and MSU crystals was associated with increased expression of TLR2 and the NO generation triggered by TLR2 signaling, inducing inflammation and cartilage deterioration. Other TLR signaling pathways producing NO release are induced by both MSU and CPPD crystals, including the PI3K/Akt/NF-κB signaling pathway and other mediators as MyD88, IRAK1, and TRAF6 [76].

4. Downregulation of immune responses by purinergic signaling

In the lymph nodes and spleen, lymphocytes are stimulated through eATP-P2X7R interaction to promote the Th1 pro-inflammatory response [77]. However, the eATP may also play an immunosuppressive role. This mechanism occurs mainly when eATP is in low (micromolar) concentrations, increasing its affinity for P2YR, located on the surface of lymphocytes. When stimulated, P2YRs promote the downregulation in the expression and release of pro-inflammatory cytokines, promoting a protective effect against excessive tissue damage [25]. Moreover, chronic exposure of DCs to low ATP doses reduces the capacity to stimulate the Th1 response, while Th2 response is favored [78], which induces the activation of T-cells with an anti-inflammatory profile. However, micromolar levels of eATP through P2Y2R induce the mechanisms of phagocytosis and increase ROS and NO production by macrophages and neutrophils [25].

The most recognized effector of anti-inflammatory responses of the purinergic system is adenosine. Extracellular adenosine is recognized by P1 receptors (A2AR and A2BR). High concentrations of adenosine activate A2AR, inhibiting the production of pro-inflammatory cytokines by macrophages [79], and also decreasing the production of ROS and NO by neutrophils, monocytes, and macrophages. However, low concentrations of adenosine (lower than micromoles) increase phagocytosis and ROS production by activation of A1R in neutrophils [25]. Also, adenosine acts on A2AR inhibiting the production of IL-12 and TNF-α in mice liver and preventing the damage by injury [10].

A1AR and A2AR are abundantly expressed at synapses in the CNS, modulating the synaptic efficacy [80]. A1AR and A2AR receptors are also expressed in the microglia and their activation promotes anti-inflammatory and migration activities, respectively [81]. In the presence of mild alterations of CNS high amounts of ATP can be release, then the activation of P2X7R induces both activation and pro-inflammatory response by microglia, leading to surrounding neuronal death [82]. Therefore, the ATP regulation in the CNS is critical; it has been suggested that CD39 expression has an essential role in cell proliferation and growth, inflammatory processes, and triggering cellular responses from ATP-induced contribute to apoptosis and host defense [83]. Moreover, *in vivo* studies on brain trauma and Alzheimer's disease, neuroinflammation has been detected associated with ATP release from microglia, occurring in an uncontrolled way mainly through pannexin/connexin hemichannels [84]. While adenosine binding to A1R or A2AR during brain disorder exerts neuroprotective and immunosuppressive capacities, respectively [85].

5. Pathogens and purinergic signaling mediating immune response

Some obligated intracellular bacterial pathogens have diverse target organs. For instance, *Mycobacterium tuberculosis* (*M. tuberculosis*) invades lungs, kidney, spinal cord, and brain, while *Chlamydia trachomatis* (*C. trachomatis*) infects genital and ocular tissue. It has been reported that these pathogens may be controlled by eATP treatment. The eATP treatment of macrophages enhances their antimicrobial properties in a P2X7R-dependent manner. For instance, eATP-related killing of *M. tuberculosis* and *C. trachomatis* within human and murine macrophages is associated with mobilization of intracellular Ca^{+2} and consequently lysosomal fusion and acidification of the containing-pathogen phagosomes [86, 87]. Moreover, adenine nucleotides (AMP and ATP) and adenosine can inhibit *C. trachomatis* growth in epithelial cells; for instance, micromolar eATP concentrations reversibly inhibit chlamydial infection via the P2X4 receptor in epithelial cells [88]. While millimolar eATP concentrations are sufficient to inhibit chlamydial infection via P2X7 receptor in macrophages [89].

Another bacteria controlled by eATP-triggered mechanisms is *P. gingivalis*, an intracellular bacterium that infects gingival epithelial cells (GECs) and the oral mucosa, causing periodontitis [90]. The mechanisms described to control *P. gingivalis* via eATP activation are (1) P2X7R-mediated apoptosis [91]; (2) ROS production via P2X7R-NADPH oxidase signaling [92] inflammasome activation and IL-1β release [93]. Conversely, adenosine-receptors signaling downregulates the immune response. For example, A2AR stimulated by its agonist CGS21680 increases the *P. gingivalis* proliferation in GECs [94]. On the other hand, *P. gingivalis* can inhibit eATP-induced apoptosis in GECs through the secretion of the enzyme nucleoside-diphosphate-kinase (NDK), which can hydrolyze eATP [92], inhibiting the three eATP-triggered mechanisms to control the bacteria.

Intracellular protozoan parasites as *Leishmania amazonensis (L. amazonensis), Toxoplasma gondii (T. gondii)*, and *Trypanosoma cruzi (T. cruzi)* may also be controlled by eATP. Murine macrophages infected with *L. amazonensis* and cells from established cutaneous lesions enhanced P2X7R expression and were more responsive to eATP activation, inhibiting parasite growth [95]. Also, UTP inhibits *L. amazonensis* infection in murine macrophages, probably by P2Y2R or P2Y4R activation, inducing morphological damage inside the parasite promoting apoptosis of macrophages, producing ROS and reactive nitrogen species (RNS), and increasing intracellular Ca^{+2} concentrations [96]. Nonetheless, several species of *Leishmania* modulate eATP and adenosine levels by directly acting on these molecules or by inducing CD39 and CD73 expression on the infected cells, influencing the immune response and contributing to parasite growth or survival [97]. First, saliva from phlebotomine sand flies, *Leishmania* promastigotes vector, is rich in adenosine and AMP, which levels are mediated by the enzymatic activity of apyrases and 5′-nucleotidases present in saliva [98]. Therefore, low levels of ATP decrease the activation of P2X and P2Y, inhibiting platelet aggregation [99], leading to the free spread of the parasite. Second, parasite-infected cells increase the expression of ectonucleotidases (CD39 and CD73) on their surface and, therefore, also the production of extracellular adenosine. Later, adenosine mediates the activation of A2BR, necessary for the expression of CD40 (DC activation marker). Conversely, the blockade of the A2BR inhibits the DC activation and interferes with T cell proliferation [100]. Third, the A2BR activation inhibits the production of NO and IL-12 by infected macrophages [101], allowing the parasite survive. Moreover, the increased A2BR expression involved IL-10 production by infected cells, in monocytes from patients with visceral leishmaniasis [102], inducing an anti-inflammatory response.

During the acute toxoplasmosis, in the mice brain, occurs an increase in purines (ATP, ADP, AMP, adenosine, xanthine, hypoxanthine, and uric acid), while in chronic toxoplasmosis reduction of the same purines, except the antioxidant, uric acid, occurs [103]. Specifically, the high levels of xanthine and hypoxanthine are associated with the inhibition of the enzyme xanthine oxidase, which catalyzes the production of uric acid, reported in *T. gondii* infected mice [104]. Moreover, in mice with toxoplasmosis, the elevated ATP levels promote increased levels of calcium inside infected cells mediated by P2X receptors, causing damage to the cells and contributing to nervous disorders and behavioral alterations [105]. Besides, *T. gondii is* eliminated by the activation of eATP-P2X7R signaling in infected macrophages mediated by the acidification of the parasitophorous vacuole and ROS production [106]. Additionally, UTP and UDP treatment in murine macrophages infected with *T. gondii* promotes 90% parasite elimination without inducing NO, ROS or apoptosis in the host cell [107]. Interestingly, UTP and UDP induced prematurely parasite egress from the host cell via P2Y2R, P2Y4R, and P2Y6R thus compromising infectivity and replication of the egressed parasites [107].

Interestingly, *T. gondii* does not produce adenosine, then the efficient transformation to the bradyzoite or long-lived cyst stage depends on the extracellular adenosine produced by ectonucleotidases expressed by infected cells [108]. CD73 expression promotes *T. gondii* differentiation and cyst formation by a mechanism dependent on adenosine generation, but independent of adenosine-receptor signaling [108]. In fact, some pathogens stimulate extracellular adenosine generation independently of the host. *Staphylococcus aureus* produces adenosine synthase A (AdsA), a cell wall-anchored enzyme which allows the bacteria to escape from clearance by phagocytosis and favoring the formation of organ abscesses. Moreover, bacteria from the gastrointestinal tract as *Enterococcus faecalis* and *Streptococcus mutans* possess homologs of adenosine synthase [109].

During acute *T. cruzi* infection in CNS, the parasite stimulates P2X7R expression in the cerebral cortex, being activated by the available eATP. As a consequence,

P2X7R activation induces ATP release, from immune and non-immune cells, chiefly via pannexin hemichannels-boosting inflammation [83]. Besides, the P2X7R activation by the parasite transialidase is involved in the massive loss of immature CD4/CD8 double-positive cells, which determine the prominent thymus atrophy in acute *T. cruzi* infection [110]. In addition, eATP mediates mechanisms to control the parasite as astrocyte proliferation and differentiation, cytokine release, and the ROS and RNS formation. Furthermore, ATP, ADP, and AMP hydrolysis occur in infected animals, related to the enzymatic modulation in the presence of high parasitism [111]. As mentioned, adenosine has a neuroprotective role; however, E-ADA activity is augmented in infected animals, producing iosine which is later used in the purine rescue pathway of *T. cruzi* [83], and in other parasites such as *Trypanosoma evansi* [112] and *Plasmodium falciparum* [113].

Interestingly, some pathogens have evolved extracellular nucleotide-hydrolyzing enzymes that mimic the ectonucleotidases expressed in the host, probably inhibiting the ATP-driven immune response [114]. For instance, the surface of *T. cruzi* expresses an Mg^{2+}-dependent ecto-ATPase enzyme (Mg-eATPase), which have higher activity in trypomastigotes, maybe promoting the host infection [115]. Another parasite with Mg-eATPase is *L. amazonensis*, whose virulent promastigotes are very efficient in hydrolyzing eATP and acquiring adenosine, which is used by the parasite [116].

The participation of the purinergic system in the immune response and pathogenesis as a consequence of SARS-CoV-2 virus infection has also been reported. SARS-CoV-2 induces the IFN response in patients, through MDA5-mediated RNA sensing with the participation of IRF3, IRF5, and NF-κB/p65 pro-inflammatory transcription factors [117]. However, Coronaviruses can evade the MDA5 recognition by forming endoplasmic reticulum-derived membrane vesicles around their RNA [118], delaying the IFN production; and in consequence, allowing higher viral replication. Viral load is highly correlated with the levels of IFNs and TNF-α, suggesting that viral load may drive high cytokine production [119]. Increased levels of TNF-α during inflammation induce ATP release via pannexin-1 channels [120]. ATP exportation out of the cell implies a deficit of intracellular ATP available for the ATP-dependent enzymes in the JAK–STAT pathway induced by IFN-I, limiting the cytokine expression and T helper cell activation [121].

At the same time, a pro-inflammatory immune response is initiated by the increase in the extracellular ATP and ADP levels in the microenvironment of immune cells activating the P2XRs and P2YRs [122]. The eATP-P2X7R signaling activation is a key process in the hyper inflammation resulting from the severe pro-inflammatory immune response against SARS-CoV-2 [123]. High levels of eATP are accompanied by the desensitization of all P1 and P2 purinergic receptors, except P2X7R, inducing more hyper inflammation [124], the worst scenario for a COVID-19 patient.

Shortly after the inflammatory explosion or simultaneously, the eATP concentration could decrease by the CD39-mediated transformation into eADP and eAMP, while adenosine quickly increases by the CD73-mediated eAMP conversion [125]. Then, immunosuppressive responses are activated by the adenosine excess in interaction with their A2AR and A2BR, including inhibition of macrophages and lymphocytes [10].

Moreover, increased eADP levels promote platelet activation and intravascular thrombosis mediated by P2YRs [126], and COVID-19 patients with pneumonia frequently developed microvascular thrombosis in their lungs [127]. In summary, the degree of involvement of purinergic receptors and their ligands in the response to SARS-CoV-2 virus infection may partially explain, the presence of asymptomatic infected people and the variation in the severity among the COVID-19 patients.

6. Perspectives

During inflammation, macrophages, NK cells, and some lymphocytes activities are impaired by the interaction of their adenosine receptors and the high extracellular levels of adenosine [10]. Therefore, the factors involved in extracellular adenosine production may be used in anti-inflammatory strategies, including the ectonucleotidases CD39 which degrades ATP into AMP, and the ectonucleotidase CD73 which converts AMP into adenosine. Following this rationale, several monoclonal antibodies (mAb) have been developed using CD73, CD39, and A2AR receptors as a target [128]. For instance, a humanized anti-CD39 mAb prevents the ATP-ADP conversion. Moreover, the enhancement of T cells and NK cells function was found, when CD39 was blocked by either antibodies or inhibitors such as POM-1; aside from increased T cell proliferation by the lack of suppression exerted by Treg cells [129].

Moreover, the prevention of AMP to adenosine conversion is also achieved using the mAb anti-CD73 which leads to its internalization [130]. As consequence, the adenosine low levels can not inhibit lymphocytes, therefore CD8 and macrophages activities are enhanced, while both myeloid suppressor cells and Treg lymphocytes are inhibited [128]. Lymphocyte proliferation is also promoted with the administration of an A2AR antagonist in two ways, removing checkpoints on both CD4+ FoxP3+ Tregs and CD8+ effector T cells development, and inhibiting the expression of the programmed death-1 receptor (PD-1) in draining lymph nodes [131]. Some of these drugs have been used as anti-cancer therapies, nevertheless, they have a potential action in many diseases based on immunosuppressive mechanisms.

On the other hand, antagonists of P2X7R as lidocaine can disrupt hyperinflammation, leading to the activation of anti-inflammatory responses. For instance, the clonal expansion of Tregs in lymph nodes is promoted by the P2X7Rs-mediated inhibition of the immune cells in the lymphatic system. Later, the Tregs control the hyperinflammation throughout their anti-inflammatory mechanisms [16]. Also, since eATP-P2Y11R signaling is highly activated in macrophages, P2Y11R antagonists maybe they can be used for the treatment of inflammatory diseases [40]. These strategies may constitute immunotherapy with promising results for inflammatory-based diseases, such as severe forms of various viral or bacterial infections, or even autoimmune diseases.

7. Conclusion

PX and PY receptors are involved in the inflammasome activation, apoptosis induction, oxidant production and activation of several immune cells, mechanisms that can control the infection of several pathogens. Conversely, adenosine is generally associated with the downregulation of inflammation. However, the effects triggered by eATP and nucleosides and their respective purinergic receptors in infected cells, depend on several aspects. These include first, the ability of the receptor expression by infected cells; second, the mechanisms to maintain the balance of nucleotide and nucleoside concentrations in the extracellular environment; and third, the survival strategies of specific pathogens.

The purinergic signaling can modulate infections by different intracellular pathogens, including viruses, bacteria, and parasites, and mediates inflammatory processes in metabolic, cardiovascular, and cancer diseases. For this reason, this knowledge field represents an important focus for future research regarding the survival and elimination of different pathogens and the maintenance of the homeostasis of the diseases related to hyper-inflammation.

Acknowledgements

Post-doctoral Fellowship of Research Support Foundation of the State of Rio de Janeiro–FAPERJ—Health Research Networks Program in the State of Rio de Janeiro—2019, Brazil. Processo E-26/ 202.139/2020.

Conflict of interest

The authors declare no conflict of interest.

Author details

Yerly Magnolia Useche Salvador
Laboratory of Thymus Research, Oswaldo Cruz Institute, Oswaldo Cruz Foundation, Rio de Janeiro, Brazil

*Address all correspondence to: yerly.useches@gmail.com

References

[1] Burnstock G. Purine and pyrimidine receptors. Cellular and Molecular Life Sciences. 2007;**64**:1471-1483

[2] Yegutkin G. Nucleotide and nucleoside converting ectoenzymes: Important modulators of purinergic signalling cascade. Biochimica et Biophysica Acta. 1783;**2008**:673-694

[3] Ishii KJ, Akira S. Potential link between the immune system and metabolism of nucleic acids. Current Opinion in Immunology. 2008;**20**:524-529. DOI: 10.1016/j.coi.2008.07.002

[4] Medzhitov R. Recognition of microorganisms and activation of the immune response. Nature. 2007;**449**:819e26

[5] Ishii KJ, Koyama S, Nakagawa A, Coban C, Akira S. Host innate immune receptors and beyond: Making sense of microbial infections. Cell Host & Microbe. 2008;**3**:352-363

[6] Idzko M, Ferrari D, Eltzschig HK. Nucleotide signalling during inflammation. Nature. 2014;**509**:310e7

[7] Fink SL, Cookson BT. Apoptosis, pyroptosis, and necrosis: Mechanistic description of dead and dying eukaryotic cells. Infection and Immunity. 2005;**73**:1907-1916. DOI: 10.1128/IAI.73.4.1907-1916.2005

[8] Di Virgilio F. Purinergic mechanism in the immune system: A signal of danger for dendritic cells. Purinergic Signalling. 2005;**1**:205-209

[9] Gessi S, Merighi S, Varani K, Cattabriga E, Benini A, Mirandola P, et al. Adenosine receptors in colon carcinoma tissues and colon tumoral cell lines: Focus on the a(3) adenosine subtype. Journal of Cellular Physiology. 2007;**211**:826-836. DOI: 10.1002/jcp.20994

[10] Ohta A, Sitkovsky M. Role of G-protein-coupled adenosine receptors in downregulation of inflammation and protection from tissue damage. Nature. 2001;**414**:916-920. DOI: 10.1038/414916a

[11] Olah ME, Stiles GL. Adenosine receptor subtypes: Characterization and therapeutic regulation. Annual Review of Pharmacology and Toxicology. 1995;**35**:581-606

[12] Coutinho-Silva R, Ojcius DM. Role of extracellular nucleotides in the immune response against intracellular bacteria and protozoan parasites. Microbes and Infection. 2012;**14**:1271e7

[13] Zimmermann H. Biochemistry, localization and functional roles of ectonucleotidases in the nervous system. Progress in Neurobiology. 1996;**49**:589-618

[14] Braun N, Sévigny J, Robson SC, Enjyoji K, Guckelberger O, Hammer K, et al. Assignment of ecto-nucleoside triphosphate diphosphohydrolase-1/cd39 expression to microglia and vasculature of the brain. The European Journal of Neuroscience. 2000;**12**:4357-4366

[15] Wang L, Olivecrona G, Gotberg M, Olsson ML, Winzell MS, Erlinge D. ADP acting on P2Y13 receptors is a negative feedback pathway for ATP release from human red blood cells. Circulation Research. 2005;**96**:189-196. DOI: 10.1161/01.res.0000153670.07559.e4

[16] Almeida-da-Silva CLC, Morandini AC, Ulrich H, Ojcius DM,

Coutinho-Silva R. Purinergic signaling during Porphyromonas gingivalis infection. Biomedical Journal. 2016;**39**:251-260. DOI: 10.1016/j.bj.2016.08.003

[17] Morandini AC, Savio LE, Coutinho-Silva R. The role of P2X7 receptor in infectious inflammatory diseases and the influence of ectonucleotidases. Biomedical Journal. 2014;**37**:169e77

[18] Coutinho-Silva R, Alves LA, de Carvalho AC, Savino W, Persechini PM. Characterization of P2Z purinergic receptors on phagocytic cells of the thymic reticulum in culture. Biochimica et Biophysica Acta. 1996;**1280**(2):217-222. DOI: 10.1016/0005-2736(95)00293-6

[19] Pelegrin P. P2X7 receptor and the NLRP3 inflammasome: Partners in crime. Biochemical Pharmacology. 2021;**187**:114385. DOI: 10.1016/j.bcp.2020.114385

[20] Yang D, He Y, Munoz-Planillo R, Liu Q, Nunez G. Caspase-11 requires the Pannexin-1 channel and the purinergic P2X7 pore to mediate pyroptosis and endotoxic shock. Immunity. 2015;**43**:923e32

[21] Lamkanfi M, Dixit VM. Modulation of inflammasome pathways by bacterial and viral pathogens. Journal of Immunology. 2011;**187**:597e602

[22] Garlanda C, Dinarello CA, Mantovani A. The interleukin-1 family: Back to the future. Immunity. 2013;**39**:1003e18

[23] Dinarello CA. Anti-inflammatory agents: Present and future. Cell. 2010;**140**:935e50

[24] Li R, Tan B, Yan Y, Ma X, Zhang N, Zhang Z, et al. Extracellular UDP and P2Y6 function as a danger signal to protect mice from vesicular stomatitis virus infection through an increase in IFN-beta production. Journal of Immunology. 2014;**193**:4515-4526. DOI: 10.4049/jimmunol.1301930

[25] Bours MJ, Swennen EL, Di Virgilio F, Cronstein BN, Dagnelie PC. Adenosine 5′-triphosphate and adenosine as endogenous signaling molecules in immunity and inflammation. Pharmacology & Therapeutics. 2006;**112**:358-404. DOI: 10.1016/j.pharmthera.2005.04.013

[26] Hasan D, Shono A, van Kalken CK, van der Spek PJ, Krenning EP, Kotani T. A novel definition and treatment of hyperinflammation in COVID-19 based on purinergic signalling. Purinergic Signal. 2021;**10**:1-47. DOI: 10.1007/s11302-021-09814-6

[27] van der Weyden L, Conigrave AD, Morris MB. Signal transduction and white cell maturation via extracellular ATP and the P2Y11 receptor. Immunology and Cell Biology. 2000;**78**:369-374. DOI: 10.1046/j.1440-1711.2000.00918.x

[28] Vaughan KR, Stokes L, Prince LR, Marriott HM, Meis S, Kassack MU, et al. Inhibition of neutrophil apoptosis by ATP is mediated by the P2Y11 receptor. Journal of Immunology. 2007;**179**:8544-8553

[29] Alkayed F, Kashimata M, Koyama N, Hayashi T, Tamura Y, Azuma Y. P2Y11 purinoceptor mediates the ATP-enhanced chemotactic response of rat neutrophils. Journal of Pharmacological Sciences. 2012;**120**:288-295

[30] Barrett MO, Sesma JI, Ball CB, Jayasekara PS, Jacobson KA, Lazarowski ER, et al. A selective high-affinity antagonist of the P2Y14 receptor inhibits UDP-glucose-stimulated

chemotaxis of human neutrophils. Molecular Pharmacology. 2013;**84**:41-49. DOI: 10.1124/mol.113.085654

[31] Arase T, Uchida H, Kajitani T, Ono M, Tamaki K, Oda H, et al. The UDP-glucose receptor P2RY14 triggers innate mucosal immunity in the female reproductive tract by inducing IL-8. Journal of Immunology. 2009;**182**:7074-7084. DOI: 10.4049/jimmunol.0900001

[32] Muller T, Fay S, Vieira RP, Karmouty-Quintana H, Cicko S, Ayata K, et al. The purinergic receptor subtype P2Y2 mediates chemotaxis of neutrophils and fibroblasts in fibrotic lung disease. Oncotarget. 2017;**8**:35962-35972. DOI: 10.18632/oncotarget.16414

[33] Oury C, Lecut C, Hego A, Wera O, Delierneux C. Purinergic control of inflammation and thrombosis: Role of P2X1 receptors. Computational and Structural Biotechnology Journal. 2015;**13**:106-110. DOI: 10.1016/j.csbj.2014.11.008

[34] Vanderstocken G, Bondue B, Horckmans M, Di Pietrantonio L, Robaye B, Boeynaems JM, et al. P2Y2 receptor regulates VCAM-1 membrane and soluble forms and eosinophil accumulation during lung inflammation. Journal of Immunology. 2010;**185**:3702-3707. DOI: 10.4049/jimmunol.0903908

[35] Muller T, Robaye B, Vieira RP, Ferrari D, Grimm M, Jakob T, et al. The purinergic receptor P2Y2 receptor mediates chemotaxis of dendritic cells and eosinophils in allergic lung inflammation. Allergy. 2010;**65**:1545-1553. DOI: 10.1111/j.1398-9995.2010.02426.x

[36] Nakano M, Ito K, Yuno T, Soma N, Aburakawa S, Kasai K, et al. UDP/P2Y6 receptor signaling regulates IgE-dependent degranulation in human basophils. Allergology International. 2017;**66**:574-580. DOI: 10.1016/j.alit.2017.02.014

[37] Banchereau J, Briere F, Caux C, Davoust J, Lebecque S, Liu YJ, et al. Immunobiology of dendritic cells. Annual Review of Immunology. 2000;**18**:767-811. DOI: 10.1146/annurev.immunol.18.1.767

[38] Schnurr M, Toy T, Stoitzner P, Cameron P, Shin A, Beecroft T, et al. ATP gradients inhibit the migratory capacity of specific human dendritic cell types: Implications for P2Y11 receptor signaling. Blood. 2003;**102**:613-620. DOI: 10.1182/blood-2002-12-3745

[39] Wilkin F, Duhant X, Bruyns C, Suarez-Huerta N, Boeynaems JM, Robaye B. The P2Y11 receptor mediates the ATP-induced maturation of human monocyte-derived dendritic cells. Journal of Immunology. 2001;**166**:7172-7177

[40] Ben Addi A, Cammarata D, Conley PB, Boeynaems JM, Robaye B. Role of the P2Y12 receptor in the modulation of murine dendritic cell function by ADP. Journal of Immunology. 2010;**185**:5900-5906. DOI: 10.4049/jimmunol.0901799

[41] Nihei OK, de Carvalho AC, Savino W, Alves LA. Pharmacologic properties of P(2Z)/P2X(7) receptor characterized in murine dendritic cells: Role on the induction of apoptosis. Blood. 2000;**96**:996-1005

[42] Wang HX, Pan W, Zheng L, et al. Thymic Epithelial Cells Contribute to Thymopoiesis and T Cell Development. Frontiers in Immunology. 2020;**10**:3099. DOI: 10.3389/fimmu.2019.03099

[43] Bisaggio RD, Nihei OK, Persechini PM, Savino W, Alves LA. Characterization of P2 receptors in

thymic epithelial cells. Cellular and Molecular Biology. 2001;**47**:19-31

[44] Woehrle T, Yip L, Elkhal A, Sumi Y, Chen Y, Yao Y, et al. Pannexin-1 hemichannel-mediated ATP release together with P2X1 and P2X4 receptors regulate T-cell activation at the immune synapse. Blood. 2010;**116**:3475-3484. DOI: 10.1182/blood-2010-04-277707

[45] Manohar M, Hirsh MI, Chen Y, Woehrle T, Karande AA, Junger WG. ATP release and autocrine signaling through P2X4 receptors regulate gammadelta T cell activation. Journal of Leukocyte Biology. 2012;**92**:787-794. DOI: 10.1189/jlb.0312121

[46] Ledderose C, Liu K, Kondo Y, Slubowski CJ, Dertnig T, Denicoló S, et al. Purinergic P2X4 receptors and mitochondrial ATP production regulate T cell migration. The Journal of Clinical Investigation. 2018;**128**:3583-3594. DOI: 10.1172/jci120972

[47] Ledderose C, Bromberger S, Slubowski CJ, Sueyoshi K, Aytan D, Shen Y, et al. The purinergic receptor P2Y11 choreographs the polarization, mitochondrial metabolism, and migration of T lymphocytes. Science Signaling. 2020;**13**:651. DOI: 10.1126/scisignal.aba3300

[48] Albayati S, Vemulapalli H, Tsygankov AY, Liverani E. P2Y(12) antagonism results in altered interactions between platelets and regulatory T cells during sepsis. Journal of Leukocyte Biology. 2020;**110**:141-153. DOI: 10.1002/jlb.3a0220-097r

[49] Sakaki H, Tsukimoto M, HaradaH, MoriyamaY, KojimaS. Autocrine regulation of macrophage activation via exocytosis of ATP and activation of P2Y11 receptor. PLoS One. 2013;**8**:e59778. DOI: 10.1371/journal.pone.0059778

[50] Guerra AN, Fisette PL, Pfeiffer ZA, Quinchia-Rios BH, Prabhu U, Aga M, et al. Purinergic receptor regulation of LPS-induced signaling and pathophysiology. Journal of Endotoxin Research. 2003;**9**:256-263. DOI: 10.1179/096805103225001468

[51] Ventura MA, Thomopoulos P. ATP and ADP activate distinct signalling pathways in human promonocyte U-937 cells differentiated with 1,25-dihydroxy-vitamin D3. Molecular Pharmacology. 1995;**47**:104-114

[52] Satonaka H, Nagata D, Takahashi M, Kiyosue A, Myojo M, Fujita D, et al. Involvement of P2Y12 receptor in vascular smooth muscle inflammatory changes via MCP-1 upregulation and monocyte adhesion. American Journal of Physiology. Heart and Circulatory Physiology. 2015;**308**:H853-H861. DOI: 10.1152/ajpheart.00862.2013

[53] Calovi S, Mut-Arbona P, Sperlágh B. Microglia and the purinergic Signaling system. Neuroscience. 2019;**405**:137-147. DOI: 10.1016/j.neuroscience.2018.12.021

[54] Choi HB, Ryu JK, Kim SU, McLarnon JG. Modulation of the purinergic P2X7 receptor attenuates lipopolysaccharide-mediated microglial activation and neuronal damage in inflamed brain. The Journal of Neuroscience. 2007;**27**:4957-4968

[55] Kobayashi K, Yamanaka H, Yanamoto F, Okubo M, Noguchi K. Multiple P2Y subtypes in spinal microglia are involved in neuropathic pain after peripheral nerve injury. Glia. 2012;**60**:1529e1539

[56] Tozaki-Saitoh H, Miyata H, Yamashita T, Matsushita K, Tsuda M, Inoue K. P2Y12 receptors in primary microglia activate nuclear factor of

activated T-cell signaling to induce C-C chemokine 3 expression. Journal of Neurochemistry. 2017;**141**:100-110. DOI: 10.1111/jnc.13968

[57] Inoue K. UDP facilitates microglial phagocytosis through P2Y6 receptors. Cell Adhesion & Migration. 2007;**1**:131-132

[58] Wen RX, Shen H, Huang SX, Wang LP, Li ZW, Peng P, et al. P2Y6 receptor inhibition aggravates ischemic brain injury by reducing microglial phagocytosis. CNS Neuroscience & Therapeutics. 2020;**26**:416-429. DOI: 10.1111/cns.13296

[59] Morioka N, Tokuhara M, Harano S, Nakamura Y, Hisaoka-Nakashima K, Nakata Y. The activation of P2Y6 receptor in cultured spinal microglia induces the production of CCL2 through the MAP kinases-NF-kappaB pathway. Neuropharmacology. 2013;**75**:116-125. DOI: 10.1016/j.neuropharm.2013.07.017

[60] Van Steenwinckel J, Reaux-Le GA, Pommier B, Mauborgne A, Dansereau MA, Kitabgi P, et al. CCL2 released from neuronal synaptic vesicles in the spinal cord is a major mediator of local inflammation and pain after peripheral nerve injury. The Journal of Neuroscience. 2011;**31**:5865e5875

[61] Rostène W, Dansereau MA, Godefroy D, Van Steenwinckel J, Reaux-Le GA, Mélik-Parsadaniantz S, et al. Neurochemokines: A menage a trois providing new insights on the functions of chemokines in the central nervous system. Journal of Neurochemistry. 2011;**118**:680e694

[62] Ting JP, Lovering RC, Alnemri ES, Bertin J, Boss JM, Davis BK, et al. The NLR gene family: A standard nomenclature. Immunity. 2008;**28**:285-287

[63] Koonin EV, Aravind L. The NACHT family - a new group of predicted NTPases implicated in apoptosis and MHC transcription activation. Trends in Biochemical Sciences. 2000;**25**:223-224

[64] Pillinger MH, Rosenthal P, Abeles AM. Hyperuricemia and gout: New insights into pathogenesis and treatment. Bulletin of the NYU Hospital for Joint Diseases. 2007;**65**:215-221

[65] Lamkanfi M, Dixit VM. Mechanisms and functions of inflammasomes. Cell. 2014;**157**:1013e22

[66] Hayyan M, Hashim MA, AlNashef IM. Superoxide ion: Generation and chemical implications. Chemical Reviews. 2016;**116**:3029-3085. DOI: 10.1021/acs.chemrev.5b00407

[67] Kim HJ, Kim CH, Ryu JH, Kim MJ, Park CY, Lee JM, et al. Reactive oxygen species induce antiviral innate immune response through IFN-λ regulation in human nasal epithelial cells. American Journal of Respiratory Cell and Molecular Biology. 2013;**49**:855-865. DOI: 10.1165/rcmb.2013-0003OC

[68] Hung SC, Choi CH, Said-Sadier N, Johnson L, Atanasova KR, Sellami H, et al. P2X4 assembles with P2X7 and pannexin-1 in gingival epithelial cells and modulates ATP-induced reactive oxygen species production and inflammasome activation. PLoS One. 2013;**8**:e70210

[69] Martinon F, Pétrilli V, Mayor A, Tardivel A, Tschopp J. Gout-associated uric acid crystals activate the NALP3 inflammasome. Nature. 2006;**440**:237-241

[70] Jin M, Yang F, Yang I, Yin Y, Luo JJ, Wang H, et al. Uric acid, hyperuricemia and vascular diseases. Frontiers in Bioscience. 2012;**1**:656-669. DOI: 10.2741/3950

[71] Liu-Bryan R, Scott P, Sydlaske A, Rose DM, Terkeltaub R. Innate immunity conferred by toll-like receptors 2 and 4 and myeloid differentiation factor 88 expression is pivotal to monosodium urate monohydrate crystal-induced inflammation. Arthritis and Rheumatism. 2005;**52**:2936-2946. DOI: 10.1002/art.21238

[72] Patterson RA, Horsley ET, Leake DS. Prooxidant and antioxidant properties of human serum ultrafiltrates toward LDL: Important role of uric acid. Journal of Lipid Research. 2003;**44**:512-521

[73] Zhang J, Diao B, Lin X, Xu J, Tang F. TLR2 and TLR4 mediate an activation of adipose tissue renin-angiotensin system induced by uric acid. Biochimie. 2019;**162**:125-133. DOI: 10.1016/j.biochi.2019.04.013

[74] Milanesi S, Verzola D, Cappadona F, Bonino B, Murugavel A, Pontremoli R, et al. Uric acid and angiotensin II additively promote inflammation and oxidative stress in human proximal tubule cells by activation of toll-like receptor 4. Journal of Cellular Physiology. 2019;**234**:10868-10876. DOI: 10.1002/jcp.27929

[75] Kocyigit I, Sener EF, Taheri S, Eroglu E, Ozturk F, Unal A, et al. Toll-like receptors in the progression of autosomal dominant polycystic kidney disease. Therapeutic Apheresis and Dialysis. 2016;**20**:615-622. DOI: 10.1111/1744-9987.12458

[76] Liu-Bryan R, Pritzker K, Firestein GS, Terkeltaub R. TLR2 signaling in chondrocytes drives calcium pyrophosphate dihydrate and monosodium urate crystal-induced nitric oxide generation. Journal of Immunology. 2005;**174**:5016-5023. DOI: 10.4049/jimmunol.174.8.5016

[77] Grassi F. The P2X7 receptor as regulator of T cell development and function. Frontiers in Immunology. 2020;**11**:1179. DOI: 10.3389/fimmu.2020.01179

[78] La Sala A, Sebastiani S, Ferrari D, Di Virgilio F, Idzko M, Norgauer J, et al. Dendritic cells exposed to extracellular adenosine triphosphate acquire the migratory properties of mature cells and show a reduced capacity to attract type 1 T lymphocytes. Blood. 2002;**99**:1715-1722

[79] Kreckler LM, Gizewski E, Wan TC, Auchampach JA. Adenosine suppresses lipopolysaccharide-induced tumor necrosis factor-alpha production by murine macrophages through a protein kinase A- and exchange protein activated by cAMP-independent signaling pathway. The Journal of Pharmacology and Experimental Therapeutics. 2009;**331**:1051-1061. DOI: 10.1124/jpet.109.157651

[80] Cunha RA. How does adenosine control neuronal dysfunction and neurodegeneration? Journal of Neurochemistry. 2016;**139**:1019-1055

[81] Luongo L, Guida F, Imperatore R, Napolitano F, Gatta L, Cristino L, et al. The A1 adenosine receptor as a new player in microglia physiology. Glia. 2014;**62**:122-132. DOI: 10.1002/glia.22592

[82] Savio LEB, de Andrade MP, da Silva CG, Coutinho-Silva R. The P2X7 receptor in inflammatory diseases: Angel or demon? Frontiers in Pharmacology. 2018;**9**:52. DOI: 10.3389/fphar.2018.00052

[83] Fracasso M, Reichert K, Bottari NB, da Silva AD, Schetinger MRC, Monteiro SG, et al. Involvement of ectonucleotidases and purinergic receptor expression during acute Chagas disease in the cortex of mice treated with resveratrol and benznidazole. Purinergic Signal.

2021;**17**:493-502. DOI: 10.1007/s11302-021-09803-9

[84] Yi C, Ezan P, Fernández P, Schmitt J, Sáez JC, Giaume C, et al. Inhibition of glial hemichannels by boldine treatment reduces neuronal suffering in a murine model of Alzheimer's disease. Glia. 2017;**65**:1607-1625

[85] Stone TW, Ceruti S, Abbracchio MP. Adenosine receptors and neurological disease: Neuroprotection and neurodegeneration. Handbook of Experimental Pharmacology. 2009;**193**:535-587. DOI: 10.1007/978-3-540-89615-9_17

[86] Kusner DJ, Adams J. ATP-induced killing of virulent mycobacterium tuberculosis within human macrophages requires phospholipase D. Journal of Immunology. 2000;**164**:379e88

[87] Kusner DJ, Barton JA. ATP stimulates human macrophages to kill intracellular virulent mycobacterium tuberculosis via calcium-dependent phagosome-lysosome fusion. Journal of Immunology. 2001;**167**:3308e15

[88] Pettengill MA, Marques-da-Silva C, Avila ML, d'Arc dos Santos OS, Lam VW, Ollawa I, et al. Reversible inhibition of chlamydia trachomatis infection in epithelial cells due to stimulation of P2X(4) receptors. Infection and Immunity. 2012;**80**:4232e8

[89] Coutinho-Silva R, Stahl L, Raymond MN, Jungas T, Verbeke P, Burnstock G, et al. Inhibition of chlamydial infectious activity due to P2X7R-dependent phospholipase D activation. Immunity. 2003;**19**:403e22

[90] Almeida-da-Silva CLC, Morandini AC, Ulrich H, Ojcius DM, Coutinho-Silva R. Purinergic signaling during Porphyromonas gingivalis infection. Biomedical Journal. 2016;**39**:251-260. DOI: 10.1016/j.bj.2016.08.003

[91] Yilmaz O, Yao L, Maeda K, Rose TM, Lewis EL, Duman M, et al. ATP scavenging by the intracellular pathogen Porphyromonas gingivalis inhibits P2X7-mediated host-cell apoptosis. Cellular Microbiology. 2008;**10**:863e75

[92] Choi CH, Spooner R, DeGuzman J, Koutouzis T, Ojcius DM, Yilmaz O. Porphyromonas gingivalis-nucleoside-diphosphate-kinase inhibits ATP-induced reactive-oxygen-species via P2X7 receptor/NADPH-oxidase signalling and contributes to persistence. Cellular Microbiology. 2013;**15**:961e76

[93] Johnson L, Atanasova KR, Bui PQ, Lee J, Hung SC, Yilmaz O, et al. Porphyromonas gingivalis attenuates ATP-mediated inflammasome activation and HMGB1 release through expression of a nucleoside-diphosphate kinase. Microbes and Infection. 2015;**17**:369e77

[94] Spooner R, De Guzman J, Lee KL, Yilmaz O. Danger signal adenosine via adenosine 2a receptor stimulates growth of Porphyromonas gingivalis in primary gingival epithelial cells. Molecular Oral Microbiology. 2014;**29**:67e78

[95] Chaves SP, Torres-Santos EC, Marques C, Figliuolo VR, Persechini PM, Coutinho-Silva R, et al. Modulation of P2X(7) purinergic receptor in macrophages by Leishmania amazonensis and its role in parasite elimination. Microbes and Infection. 2009;**11**:842e9

[96] Marques-da-Silva C, Chaves MM, Chaves SP, Figliuolo VR, Meyer-Fernandes JR, Corte-Real S, et al. Infection with Leishmania amazonensis upregulates purinergic receptor expression and induces host-cell susceptibility to UTP-mediated

apoptosis. Cellular Microbiology. 2011;**13**:1410e28

[97] Figueiredo AB, Souza-Testasicca MC, Afonso LCC. Purinergic signaling and infection by Leishmania: A new approach to evasion of the immune response. Biomedical Journal. 2016;**39**:244-250. DOI: 10.1016/j.bj.2016.08.004

[98] Katz O, Waitumbi JN, Zer R, Warburg A. Adenosine, AMP, and protein phosphatase activity in sandfly saliva. The American Journal of Tropical Medicine and Hygiene. 2000;**62**:145e50

[99] Vial C, Rolf MG, Mahaut-Smith MP, Evans RJ. A study of P2X1 receptor function in murine megakaryocytes and human platelets reveals synergy with P2Y receptors. British Journal of Pharmacology. 2002;**135**:363e72

[100] Figueiredo AB, Serafim TD, Marques-da-Silva EA, Meyer-Fernandes JR, Afonso LC. Leishmania amazonensis impairs DC function by inhibiting CD40 expression via A2B adenosine receptor activation. European Journal of Immunology. 2012;**42**:1203e15

[101] Gomes RS, de Carvalho LC, de Souza VR, Fietto JL, Afonso LC. E-NTPDase (ecto-nucleoside triphosphate diphosphohydrolase) of Leishmania amazonensis inhibits macrophage activation. Microbes and Infection. 2015;**17**:295e303

[102] Amit A, Kumar S, Dikhit MR, Jha PK, Singh AK, et al. Up regulation of A2B adenosine receptor on monocytes are crucially required for immune pathogenicity in Indian patients exposed to Leishmania donovani. Cytokine. 2016;**79**:38e44

[103] Tonin AA, Da Silva AS, Casali EA, Silveira SS, Moritz CE, Camillo G, et al. Influence of infection by toxoplasma gondii on purine levels and E-ADA activity in the brain of mice experimentally infected mice. Experimental Parasitology. 2014;**142**:51-58. DOI: 10.1016/j.exppara.2014. 04.008

[104] Gherardi A, Sarciron ME, Francoise A, Peyron F. Purine pathway enzymes in a cyst forming strain of toxoplasma gondii. Life sciences. 1999;**65**:1733-1738

[105] Edwards FA, Gibb AJ, Colquhoun D. ATP receptor-mediated synaptic currents in the central nervous system. Nature. 1992;**359**:144-147

[106] Correa G, da SC M, de Abreu Moreira-Souza AC, Vommaro RC, Coutinho-Silva R. Activation of the P2X(7) receptor triggers the elimination of toxoplasma gondii tachyzoites from infected macrophages. Microbes and Infection. 2010;**12**:497e504

[107] Moreira-Souza AC, Marinho Y, Correa G, Santoro GF, Coutinho CM, Vommaro RC, et al. Pyrimidinergic receptor activation controls toxoplasma gondii infection in macrophages. PLoS One. 2015;**10**:e0133502

[108] Mahamed DA, Mills JH, Egan EE, Denkers EY, Bynoe MS. CD73-generated adenosine facilitates toxoplasma gondii differentiation to long-lived tissue cysts in the central nervous system. Proceedings. National Academy of Sciences. United States of America. 2012;**109**:16312-16317

[109] Thammavongsa V, Kern JW, Missiakas DM, Schneewind O. Staphylococcus aureus synthesizes adenosine to escape host immune responses. The Journal of Experimental Medicine. 2009;**206**:2417e27

[110] Henriques-Pons A, DeMeis J, Cotta-De-Almeida V, Savino W,

Araújo-Jorge TC. Fas and perforin are not required for thymus atrophy induced by Trypanosoma cruzi infection. Experimental Parasitology. 2004;**107**:1-4. DOI: 10.1016/j.exppara.2004.04.010

[111] Santos EC, Novaes RD, Cardoso SA. Oliveira LL implication of purinergic signalling pathways in clinical management of Chagas disease. OA Biotechnology. 2013;**2**:27

[112] Da Silva AS, Bellé LP, Bitencourt PE, Perez HA, Thomé GR, Costa MM, et al. Trypanosoma evansi: Adenosine deaminase activity in the brain of infected rats. Experimental Parasitology. 2011;**127**:173-177. DOI: 10.1016/j.exppara.2010.07.010

[113] Ivanov A, Matsumura I. The adenosine deaminases of plasmodium vivax and plasmodium falciparum exhibit surprising differences in ligand specificity. Journal of Molecular Graphics & Modelling. 2012;**35**:43-48. DOI: 10.1016/j.jmgm.2012.2.004

[114] Almeida-da-Silva CLC, Morandini AC, Ulrich H, Ojcius DM, Coutinho-Silva R. Purinergic signaling during Porphyromonas gingivalis infection. Biomedical Journal. 2016;**39**:251-260. DOI: 10.1016/j.bj.2016.08.003

[115] Bisaggio DF, Peres-Sampaio CE, Meyer-Fernandes JR, Souto-Padron T. Ecto-ATPase activity on the surface of Trypanosoma cruzi and its possible role in the parasite-host cell interaction. Parasitology Research. 2003;**91**:273e82

[116] Berredo-Pinho M, Peres-Sampaio CE, Chrispim PP, Belmont-Firpo R, Lemos AP, Martiny A, et al. A Mg-dependent ecto-ATPase in Leishmania amazonensis and its possible role in adenosine acquisition and virulence. Archives of Biochemistry and Biophysics. 2001;**391**:16e24

[117] Yin X, Riva L, Pu Y, Martin-Sancho L, Kanamune J, Yamamoto Y, et al. MDA5 governs the innate immune response to SARS-CoV-2 in lung epithelial cells. Cell Reports. 2021;**34**:108628. DOI: 10.1016/j.celrep.2020.108628

[118] Kindler E, Thiel V. To sense or not to sense viral RNA–essentials of coronavirus innate immune evasion. Current Opinion in Microbiology. 2014;**20**:69-75. DOI: 10.1016/j.mib.2014.05.005

[119] Lucas C, Wong P, Klein J, Castro TBR, Silva J, Sundaram M, et al. Longitudinal analyses reveal immunological misfiring in severe COVID-19. Nature. 2020;**584**:463-469

[120] Lohman AW, Leskov IL, Butcher JT, Johnstone SR, Stokes TA, Begandt D, et al. Pannexin 1 channels regulate leukocyte emigration through the venous endothelium during acute inflammation. Nature Communications. 2015;**6**:7965. DOI: 10.1038/ncomms8965

[121] Seif F, Khoshmirsafa M, Aazami H, Mohsenzadegan M, Sedighi G, Bahar M. The role of JAK-STAT signaling pathway and its regulators in the fate of T helper cells. Cell Communication and Signaling: CCS. 2017;**15**:23. DOI: 10.1186/s12964-017-0177-y

[122] Eltzschig HK, Sitkovsky MV, Robson SC. Purinergic signaling during inflammation. The New England Journal of Medicine. 2012;**367**:2322-2333

[123] Di Virgilio F, Tang Y, Sarti AC, Rossato M. A rationale for targeting the P2X7 receptor in coronavirus disease 19 (Covid-19). British Journal of Pharmacology. 2020;**177**:4990-4994. DOI: 10.1111/bph.15138

[124] Klaasse EC, Ijzerman AP, de Grip WJ, Beukers MW. Internalization

and desensitization of adenosine receptors. Purinergic Signal. 2008;**4**: 21-37. DOI: 10.1007/s11302-007-9086-7

[125] Abdel-Magid AF. Inhibitors of CD73 may provide a treatment for Cancer and autoimmune diseases. ACS Medicinal Chemistry Letters. 2017;**8**:781-782. DOI: 10.1021/acsmedchemlett.7b00255

[126] Nylander S, Mattsson C, Ramstrom S, Lindahl TL. Synergistic action between inhibition of P2Y12/P2Y1 and P2Y12/thrombin in ADP- and thrombin-induced human platelet activation. British Journal of Pharmacology. 2004;**142**:1325-1331. DOI: 10.1038/sj.bjp.0705885

[127] McGonagle D, O'Donnell JS, Sharif K, Emery P, Bridgewood C. Immune mechanisms of pulmonary intravascular coagulopathy in COVID-19 pneumonia. The Lancet Rheumatology. 2020;**2**:e437-e445. DOI: 10.1016/s2665-9913(20)30121-1

[128] Abouelkhair MA. Targeting adenosinergic pathway and adenosine A2A receptor signaling for the treatment of COVID-19: A hypothesis. Medical Hypotheses. 2020;**144**:110012. DOI: 10.1016/j.mehy.2020.110012

[129] Vigano S, Alatzoglou D, Irving M, Ménétrier-Caux C, Caux C, Romero P. Targeting adenosine in cancer immunotherapy to enhance T-cell function. Frontiers in Immunology. 2019;**10**:925

[130] Terp MG, Olesen KA, Arnspang EC, Lund RR, Lagerholm BC, Ditzel HJ. Anti-human CD73 monoclonal antibody inhibits metastasis formation in human breast cancer by inducing clustering and internalization of CD73 expressed on the surface of cancer cells. Journal of Immunology. 2013;**191**:4165-4173

[131] Leone RD, Lo Y-C, Powell JD. A2aR antagonists: Next generation checkpoint blockade for cancer immunotherapy. Computational and Structural Biotechnology Journal. 2015;**13**:265-272

Chapter 2

Cross Talk of Purinergic and Immune Signaling: Implication in Inflammatory and Pathogenic Diseases

Richa Rai

Abstract

Purine derivatives like adenosine 5′-triphosphate (ATP) is the powerhouse of the cell and is essential to maintain the cellular homeostasis and activity. Besides this they also act as a chemical messenger when released into the extracellular milieu because of stress and cellular insult. The extracellular ATP (eATP) as well as its metabolite adenosine triggers purinergic signaling affecting various cellular processes such as cytokine and chemokine production, immune cell function, differentiation, and maturation, and mediates inflammatory activity. Aberrant purinergic signaling had been implicated in several diseased conditions. This chapter will focus on the dynamics of purinergic signaling and immune signaling in driving under various diseased conditions like autoimmunity and infectious disease.

Keywords: ATP, adenosine, ectonucleotidases, CD39, CD73, purinergic signaling, systemic lupus erythematosus, rheumatoid arthritis, infectious disease, SARS-CoV2

1. Introduction

Adenosine 5′-triphosphate (ATP) is abundantly generated in the cytosol through respiration and glycolysis. Primarily, these are the "energy currency" of the cell as ATP hydrolysis release energy and is essential to maintain the cellular homeostasis and activity [1]. Extracellular activity of ATP was first described by Drury and Szent-Györgyi in 1929 [2]. Later, in 1970s, ATP was shown to be involved in non-adrenergic, non-cholinergic nerve-mediated responses, and further its function as a neurotransmitter was established that led to introduction of the term "purinergic signaling" [3]. Burnstock has described about the purinergic signaling and purinergic systems in very detail, which consists of (a) purine or pyrimidine derivatives that serve as an "extracellular messenger," (b) "membrane transporter" that are responsible for the extracellular release of these nucleotides or nucleosides, (c) "metabolizing enzymes" present on the cell surface that hydrolyze the ATP to adenosine diphosphate (ADP) then to adenosine monophosphate (AMP) and adenosine and, (d) "purinergic receptors" that sense the extracellular purine or pyrimidine derivatives [3–7].

In the beginning, purinergic signaling was determined to have a role in neuronal signaling but now, their role in immune responses, inflammation, pain, exocrine and endocrine secretion, platelet aggregation, and endothelial-mediated vasodilatation had been explored and established [3, 4, 6, 8]. Additionally, cross talk of purinergic signaling with other signaling network also associates with the impact on cell proliferation, differentiation, and death that occur during the development and regeneration processes. Under normal condition, purinergic signaling operate in a very well-regulated manner to maintain the physiological function of different organ systems. Dysregulation in any component of the purinergic signaling network depending on the expression or activation of purinergic receptors, ectonucleotidases or release of agonist from damaged cell resulting from stress, inflammation serves as a potent modulator of inflammation and key promoters of host defenses, immune cells activation, pathogen clearance, and tissue repair that contributes to the disease pathogenesis [9]. Thus, their knowledge is of great importance for a full understanding of the pathophysiology of acute and chronic inflammatory diseases and will give an insight on novel therapeutic approaches to overcome inflammation. This chapter describes the component of purinergic system, its cross talk with immune signaling. Major focus of this chapter is to present the dynamics of purinergic signaling under normal physiological condition and its role in modulating the immune and inflammatory response under various diseased conditions like autoimmunity, and microbial infection.

2. Purinergic system and its component on immune cells: an immunomodulator

2.1 Mechanism of release of nucleotides

Extracellular ATP (eATP) has been well established as a ligand for autocrine and paracrine signaling that has a pathophysiological role. In addition, to eATP other nucleotides and nucleosides such as the adenosine, adenosine monophosphate (AMP), adenosine diphosphate (ADP), uridine diphosphate (UDP), uridine triphosphate (UTP), and nicotinamide adenine dinucleotide (NAD+) also serve as a potent purinergic signaling modulator. The release of nucleotides into the extracellular space occurs via regulated and unregulated mechanisms. Regulated mode of release of nucleotide is mediated through classical exocytosis [10] or conductive ATP release through ATP-permeable channels [11]. Currently, five groups of ATP-release channels are known such as: connexin hemichannels, Pannexin (PANX), calcium homeostasis modulator 1 (CALHM1), volume-regulated anion channels (VRACs), and maxi-anion channels (MACs) [12].

The ATP release by exocytosis is an active release mechanism that involves vesicular nucleotide transporter (VNUT). It is responsible for the accumulation and exocytosis of ATP from exocytotic vesicles that occurs in a proton-dependent electrochemical gradient manner generated by a vacuolar-ATPase (v-ATPase). Further, intracellular Ca^{+2} level and soluble N-ethylmaleimide-sensitive factor attachment protein receptor (SNARE) drives the fusion of the exocytotic vesicles with the plasma membrane ultimately resulting in the release of nucleotides into the extracellular space [13, 14]. Hemichannels are the ATP permeable channels that support the release of ATP under specific pathological condition. Primarily, these channels contribute to various cellular and physiological functions by forming gap junctions or hemichannels, to allow intercellular communication. They are categorized into two based on their functions, [1] connexins that has both gap junction and form hemichannel function whereas [2] PANX only form hemichannel [15]. These channels are in closed state under normal condition to avoid the loss of vital ionic, energetic, and metabolic gradients. However, chemical and biochemical stimuli resulting from pathological conditions trigger their opening and lead to release of ATP. Till date, 21 isoforms of connexins are reported in human of which connexin-43, -37, -26, and -36 have been shown to support ATP release [16]. Connexin-43 is widely expressed and very well studied. It is activated by increase in the intracellular Ca^{+2} concentration, plasma membrane depolarization, reactive oxygen species (ROS) or nitric oxide (NO) [17, 18]. The PANX (PANX) family is comprised of three members, PANX-1, -2, and -3 of which PANX-1 and -3 are widely expressed in different tissues, while PANX-2 is exclusively found in the brain [19]. In resting state, PANX channels are closed, mainly due to the blockage of the pore by C-terminal tail from the intracellular side [18]. However, in response to apoptosis or pyroptosis, C-terminal tail gets cleaved by caspase-3, -7, or -11 leading to opening of PANX-1 and allows nucleotides to cross the plasma membrane [20, 21]. Additionally, other stimuli such as intracellular calcium increase, redox potential changes, mechanical stress, and activation of the P2X7R can trigger PANX-1 channel opening [22].

Another mechanism involves the disruption of the cell membrane by apoptosis, necrosis, pyroptosis, or netosis, which leads to the unregulated leakage of ATP as well as other large cytosolic molecules including enzymes [11, 23, 24].

2.2 Metabolism of extracellular nucleotides and nucleosides

The life span of eATP is controlled by purinergic ectoenzymes that coordinate a sequential two-step process of hydrolyzing ATP into AMP and then into the potent anti-inflammatory adenosine. This make ectonucleotidases enzymes a crucial component of the purinergic system, which balance the level of eATP as well as other nucleotide derivatives UTP, NAD+, and their metabolites, thereby controlling the activation of purinergic receptors and biochemical composition of the inflammatory microenvironment. These enzymes are classified into four major families: (a) *ectonucleoside triphosphate phosphohydrolases (NTPDases)*: This group of enzymes are further classified into 8 subfamilies—NTPDase 1 (CD39), 2, 3, and 8, which are expressed on the cell surface, whereas NTPDases 4–7 are present in the intracellular organelles. Of these 8 subfamilies, NTPDase1 (CD39) is the best characterized that hydrolyses ATP to ADP and further to AMP. CD39 is expressed on wide variety of immune cell, e.g., monocytes, dendritic cells (DCs), T regulatory (Treg) cells, and natural killer (NK) cells. (b) *nicotinamide adenine dinucleotide glycohydrolase (NAD glycohydrolase/CD38)*: CD38 is a cell surface glycoprotein highly expressed in hematopoietic tissues such as the bone barrow and lymph nodes. Among immune cells, CD38 is highly expressed on monocytes, macrophages, DCs, neutrophils, innate lymphoid cells (ILC), NK cells, T and B cells. It hydrolyses NAD+ to cyclic-ADP ribose (cADPR) and then to AMP. (c) *ecto-5′-nucleotidase (NT5E/CD73)*: CD73 degrades AMP generated by CD39 or CD38 to adenosine. It is expressed on stromal cells, follicular DCs, endothelial cells, neutrophils, macrophages, and subpopulations of T cells. and (d) *ectonucleotide pyrophosphatase/phosphodiesterase (NPPs)*: NPPs include 7 members NPP 1–7. NPP1–3 degrade nucleoside triphosphates and diphosphates, NAD+, UDP-sugars, and dinucleoside polyphosphates. NPP2 also known as autotaxin (ATX) has unique property of hydrolyzing nucleotide as well as phospholipids but acts more efficiently on later to generate the bioactive phospholipid mediator's lysophosphatidic acid (LPA) and sphingosine-1-phosphate (S1P). NPP6 and 7 hydrolyzes phospholipids only, whereas catalytic properties of NPP4 and 5 are not known. Some NPPs are expressed on liver and intestinal epithelia, neuronal cells; NPP1 is also expressed on B and T cells [25–28]. The ectonucleotidases are present on almost all types of immune cells, but their expression pattern changes in a function dependent manner and controls the pro-inflammatory and anti-inflammatory condition to avoid any pathological conditions like autoimmunity, cancer, and infectious disease.

Briefly, CD39 has an anti-inflammatory property that controls the extracellular level of ATP by converting it into adenosine in conjunction with CD73. CD39 and CD73 exhibit an immunosuppressive activity as shown by its expression on Tregs cells [29–31]; CD8 T cells [32] and B cells [33] and inhibits the pathogenic T cells. Breakdown of eATP by CD39 prevents the activation of P2X7R and attenuates the secretion of IL-1β and IL-18 [34]. The expression pattern of CD38 varies during the differentiation and maturation of B and T cells [35, 36]. The enzymatic activity of CD38 generates cADPR/ADPR and triggers Ca^{+2} release from intracellular stores and Ca^{+2} influx from the extracellular space that have role in transmigration and chemotaxis of neutrophils, monocytes and DCs, and cytokine release [37]. Elevated level of cADPR/ADPR and intracellular Ca^{+2} regulates cellular chemotaxis [38], phagocytosis [39], and antigen presentation [40] in a CD38 dependent manner. Thus, dysregulation of CD38 has been implicated in several inflammatory pathologies such as autoimmunity and cancer [37, 41]. It is important to note that cADPR is generated by hydrolysis of NAD+, disruption in the metabolism of NAD+ has been associated with multiple pathological conditions [42]. Different types of NPPs have been implicated in a various of pathologic conditions such as tumor invasion and metastasis, inflammation, and angiogenesis (NPP2), tissue calcification and bone development (NPP1), and hemostasis and platelet aggregation (NPP4) [43]. However, NPP2 (ATX) is widely studied, ATX-LPA signaling axis induces inflammatory mediators such as IL-8, IL-6, TNF-α, and growth factors such as the vascular endothelial growth factor (VEGF) and the granulocyte colony-stimulating factor (G-CSF) thereby augmenting the cytokine production and lymphocyte infiltration that ultimately aggravates the inflammation in conditions such as asthma, pulmonary fibrosis, and rheumatoid arthritis [44, 45].

2.3 Purinergic receptors

Purinergic receptors are divided into two subtypes based on their binding tendency to different purine derivatives—P1 receptor (P1R) has affinity to bind adenosine only, whereas P2 can bind ATP, ADP, UDP-glucose, UDP and UTP [46]. Adenosine receptors (AR) belong to rhodopsin-like family of G protein receptors and consist of four subtypes such as A1, A2A, A2B, and A3. Adenosine generated by the hydrolysis of extracellular ATP, ADP, or AMP are either metabolized by adenosine deaminase (ADA) or shuttled back to the cells via two types of transporters, the equilibrate nucleoside transporters (ENTs) and the concentrative nucleoside transporters (CNTs) to stimulate various intracellular pathways like AMP-activated protein kinase, adenosine kinase and S-adenosyl homocysteine hydrolase [47]. Although it may depend on the concentration of adenosine and the given P1 receptor subtype engaged, but adenosine primarily, have anti-inflammatory and immune suppressive functions. The immunosuppressant activity of adenosine relies on the inhibition of virtually all immune cell populations such as T and B lymphocytes, NK cells, DCs, granulocytes, monocytes, and macrophages.

P2 receptors further categorized into two families based on molecular structure and second messenger systems, namely P2X ionotropic ligand-gated ion channel receptors that only binds to ATP and P2Y metabotropic G protein-coupled receptors (GPCR) can bind to ADP, UDP- glucose, UDP, and UTP [46]. The family of P2X receptors comprises seven members (P2X1–7), which perform tissue-specific functions by forming homo- or hetero-trimeric complexes. At least three P2X subunits assemble to form hetero- (e.g., P2X2/3 and P2X1/5) or homo-trimeric (P2X7) channels. This kind of assembly confers to P2X receptors a large repertoire of physiological functions in different tissues. Among P2XRs, the P2X7R has a special place in inflammation since its stimulation promotes NLRP3 inflammasome assembly and the associated IL-1β secretion. There are eight subtypes of P2Y receptors, which is further characterized into two subfamilies P2Y1 and P2Y12 based on their coupling to G_q and G_i, respectively. P2Y1 subfamily includes P2Y1, P2Y2, P2Y4, P2Y6, and P2Y11 receptors. The second subfamily is P2Y12, which contains P2Y12, P2Y13, and P2Y14 receptors. Each P2Y receptors has different affinity towards different nucleotides and has a tissue-specific function. For instance, P2YR11 has affinity for ATP; P2YR1, P2YR12, and P2YR13 for ADP; P2YR2 and P2YR4 for UTP; P2YR6 for UDP; and P2YR14 for UDP-glucose and UDP-galactose [5].

The purinoceptors are expressed on almost all kinds of peripheral tissues and are involved in short-term as well as long-term regulation of variety of functions, ranging from neuromuscular and synaptic transmission to secretion in gut, kidney, liver, and reproductive systems. Their contribution in immune signaling is enormous, as these receptors are expressed on almost all types of immune cells. The purine nucleotides orchestrate the onset, magnitude duration, and resolution of the inflammatory response through the activation of purinergic receptors, which is also governed by the activity of ectonucleotidases (**Figures 1** and **2**). Any alterations in the purinergic machinery could contribute to the pathophysiological processes underlying the onset

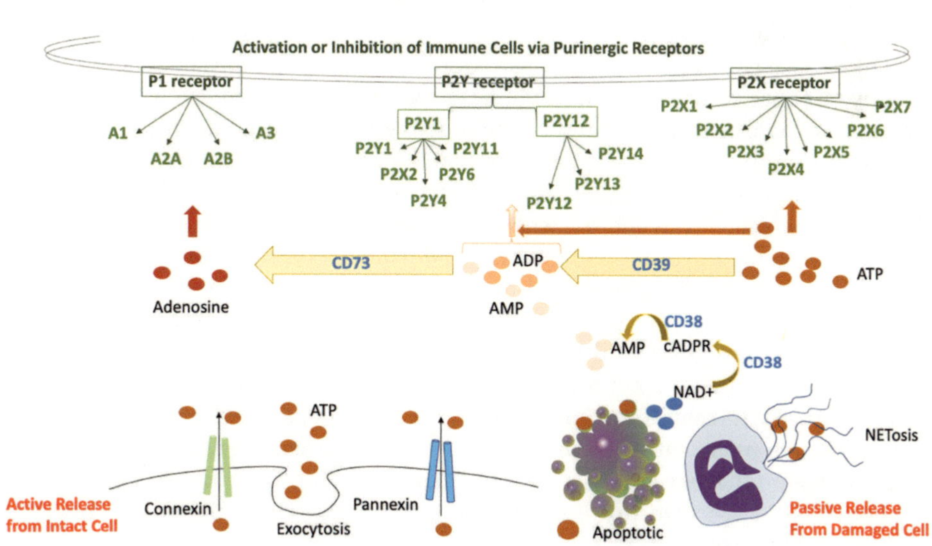

Figure 1.
Cartoon depicting the components of purinergic system and their functions. Mechanism of nucleotide release from the intact cells via exocytosis or transport channels as well as leakage of ATP from apoptotic and netosis (bottom of the image). The nucleotide triggers the activation of immune cells via specific purinergic receptor (top of the image). Activation of purinergic receptors ATP or their hydrolyzed metabolites ADP/AMP and adenosine by ectonucleotidases (middle of the image).

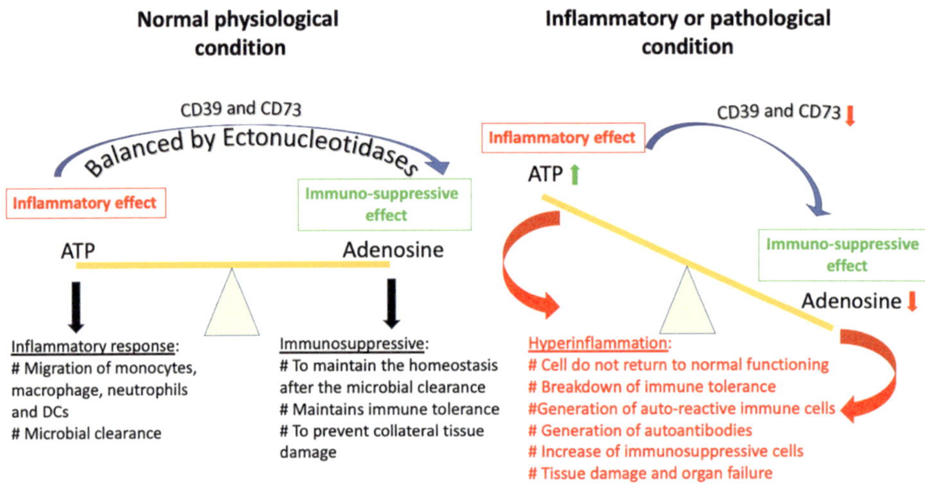

Figure 2.
Pictorial representation of cross talk of purinergic and immune signaling during normal physiological condition and inflammatory condition.

and development of immunological diseases, neurodegeneration, cancer, diabetes, and hypertension [7, 46, 48].

3. Interplay of purinergic signaling and immune signaling on inflammatory response

Beyond the physical and chemical barrier of skin and mucous lining, our body is guarded from the pathogens as well as self-attacking/cancerous cells by two different kinds of immune responses that acts in a coordinated manner. This includes (a) innate immune response comprising myeloid lineage derived cells (monocytes, macrophages, neutrophils, and DCs) and NK cells derived from lymphoid progenitors, and (b) adaptive immune response consists of B and T cells. Innate immune response provides the first line of defense against pathogens. It is an antigen-independent defense mechanism that is elicited when immune cells encounter pathogens. This response has no memory and remains similar during the lifetime. On the other hand, adaptive immune response is an antigen dependent, antigen specific, and has the tendency to form memory cells to elicit rapid response based on the previous encounter with the similar kind of antigen or pathogenic exposure. Innate and adaptive immune responses are not mutually exclusive defense mechanisms. They work in a very organized fashion and complement the functions, as activation of T cells requires antigen presentation by professional antigen presenting cells (dendritic cells, B-cells, or macrophages), together with the major histocompatibility complex (MHC) type I or II [49]. Defects in any of the component increases the vulnerability towards infection and disease.

ATP and adenosine are the key modulators of the immune response, ATP being an immunostimulant, whereas adenosine has an immunosuppressive effect thus balance between the two is crucial for the proper functioning of immune system. Extracellular signals by ATP and adenosine are detected and transduced by P2 and P1 receptors (**Figure 2**), respectively which is present on all kinds of immune cells, thus purinergic signaling affects all aspects of immunity and inflammation [50], which is described in detail in further section.

3.1 Effect on innate immune signaling

3.1.1 Monocytes/macrophages

Macrophages are the subset of myeloid cells that have immune surveillance function and sense even a minute changes in the tissue microenvironment. They express a variety of pattern recognition receptors (PRRs) that are present either on the surface, cytosol or in the endosome such as toll like receptors (TLRs), NOD like receptor (NLRs), retinoic acid inducible gene I like receptors (RLR), transmembrane C-type lectin receptors, and absent in melanoma (AIM)2-like receptors (ALRs) that recognize either pathogen associated or damage associated molecular patterns (PAMP and DAMP, respectively). These cells are highly plastic that could undergo profound metabolic modifications after sensing the pathogens or damage signal via PRRs to elicit the immune response. In addition, macrophages are endowed with purinergic P1, P2X, and P2Y receptors that also respond to damage associated molecules, extracellular nucleotides, and their derivatives, and undergo reprogramming from pro-inflammatory profile M1-like phenotype to an anti-inflammatory M2-like phenotype. As indicated by Elliott et al., that monocytes or macrophages sense extracellular nucleotide as a danger signal for "find me" or "eat me" to engulf and phagocytose the dying cells [24]. These cells not only sense the distant signal but also amplify the signaling for chemotaxis by releasing ATP by "autocrine purinergic loop" via P2Y2 and A3 receptors [51, 52]. Macrophages control their activation state in an autoregulatory mechanism by inducing the production of ATP and extracellular degradation to adenosine. Deficiency of CD39 promotes a sustained inflammatory activation state and inhibits the switch to an immunosuppressive phenotype [53]. Presence of extracellular adenosine stimuli in macrophages drives the polarization towards M2 phenotype with diminished expression of inflammatory genes TNF-α and IL-6 and increased expression of anti-inflammatory cytokines such as IL-10 and VEGF via A2A and A2B receptors [54]. Furthermore, macrophages exhibit a unique repertoire of P2X receptors such as expression of P2X1, P2X4, as well as P2X7 [55]. Among P2Y receptors, P2Y1 and P2Y4 receptors play minor roles, whereas the functions of P2Y2, P2Y6, P2Y11, P2Y12, P2Y13, and P2Y14 are more established in the macrophage biology as described elsewhere [56]. A recent study demonstrated that bone marrow derived macrophages display unique expression pattern of purinergic receptors that correlates with a M1or M2 inflammatory phenotype. M1 phenotype exhibit a unique and more pronounced P2X7 negative macrophage population, which associates with decreased inflammasome formation. P1 receptors A2A and A2B are upregulated in M1 and M2. P2Y1 and P2Y6 exclusively upregulated in M2, whereas P2Y13 and P2Y14 are overexpressed in M1 [57]. This unique feature demonstrates capability of purinergic receptors on macrophages to adapt to pro- and anti-inflammatory macrophage differentiation with functional consequences to nucleotide stimulation.

3.1.2 Dendritic cells

DCs are professional antigen-presenting cells (APCs), which has a crucial role in initiating and regulating the adaptive immune response by directing the activation and differentiation of naive T cells. Immature DCs (iDCs) sense the danger signals in the similar fashion as monocytes and macrophages do, however upon exposure, DCs lose their phagocytotic capacity, migrate to secondary lymphoid organs and transition to a mature DC (mDC) by acquiring MHC and costimulatory molecules, such as CD54, CD80, CD83, and CD86. Migration of DCs to the inflamed tissue is mediated by A1 and A3 ARs [58]. Adenosine upregulates the expression of co-stimulatory molecules on mDCs [59]. Both, A2A and A2B ARs suppress maturation of DCs as well as their capacity to initiate Th1 response, however, it increases pro-angiogenic VEGF, IL-10 and cytokines that contribute to Th17 cell polarization [59, 60]. Adenosine also mediates the attraction of DC and Treg cells, which is crucial for the immunosuppressive activity of Treg cells [61]. Similarly, ATP also acts as a chemoattractant for iDCs, and enhance the migration by autocrine signaling loop mechanism via P2X7. This signaling is further amplified by the release of ATP by PANX-1 channels [62]. Furthermore, eATP had been shown to activate P2X7R to promote the maturation of dendritic cells via NF-κB (p65) pathway [63]. On the other hand, P2Y6 has inhibitory role in the maturation and activation of DCs via NF-κB by inhibiting the production of IL-12 and IL-23 and the polarization of Th1 and Th17. Loss of P2Y6 enhances the DC mediates differentiation of Th1 and Th17 subsets [64]. The ATP-P2X7 signaling axis of DCs also promotes interleukin (IL)-1β and IL-18 secretion by activating NLRP3 inflammasome and induces Th2/Th17 differentiation [65]. P2X4 acts in conjunction with P2X7 to regulate IL-1β production by DCs [66]. As described previously, the balance of proinflammatory-ATP and anti-inflammatory adenosine is regulated by CD39 and CD73 present on the immune cells. In context of DCs, their expression fine tunes the DCs function either as tolerance (higher expression) or as immunity (lower expression) ensues [67, 68].

3.1.3 Neutrophils

Neutrophils belongs to the granulocyte family, which has a major role during the early stages of the inflammatory response. They are the first cell to arrive at the inflammation site, which employ an extracellular ATP-dependent mechanism to generate a chemotactic gradient and orientate its migration. Remarkably, the purinergic system regulates many effector functions of neutrophils such as phagocytosis, oxidative burst, degranulation, and neutrophil extracellular traps (NETs) formation via Netosis [69, 70]. Apoptotic neutrophils release ATP to stimulate mononuclear phagocytic cell influx and promote engulfment and clearance functions. Nucleotides released as a result of the apoptosis and netosis serve as danger or find me signal to initiate immune cell chemotaxis via P2Y2 receptor towards inflamed tissue and fine-tuned control local inflammation and promote phagocytosis and clearance [24, 71]. Similar to other phagocytic cells such as monocytic and dendritic cells, neutrophils in the immune microenvironment also release ATP via PANX-1 to induce chemotaxis by autocrine stimulation of P2Y2 [51, 52, 62, 72]. On the other hand, P1 receptor, A2A (activated by adenosine) blocks the chemoattractant signaling, whereas alternative binding of adenosine to A3 receptors, stimulate immune migration. Thus, P2Y2 and A3 receptors are responsible for the amplification of the chemotaxis signal via feedback loop mechanism [52]. P2Y2 receptors play crucial role in neutrophil activation by regulating the release of IL-8, a major chemokine for neutrophils chemotaxis [73]. IL-8 secretion is in turn controlled by CD39 [74]. Thus, the local microenvironment composed by ATP and the consequent degradation to adenosine by CD39 and CD73 ectoenzymes influence reprogramming of the innate immune cells and their response towards pathogens and other diseased condition.

3.1.4 Natural killer cells

NK cells are considered as a component of innate immune system due to the lack antigen-specific cell surface receptors but morphologically they resemble lymphocytes as they originate from the common lymphoid progenitor cell in the bone marrow. NK cells exert sophisticated biological functions that attribute to both innate and adaptive immunity, thus the functional boundary between these two arms of the immune response is obscure [75]. These cells express a repertoire of activating (NKG2 C-H) and inhibitory receptors (NKG2 A and B) through which it interacts with pathogens by recognizing MHC-I molecule [76]. Activation of the NK cell leads to cytolytic killing of infected cells. Adenosine receptors A1, A3 and A2A, A2B have an antagonistic effect in controlling the intracellular cAMP levels. A1, A3 inhibits the adenylyl cyclase and decreases the intracellular cAMP level, which has a stimulatory effect on NK cell and promote the cytotoxic activity whereas, A2A and A2B has the immunosuppressive effects on NK cells [76, 77]. NAD+ and ADP-ribose inhibited human NK proliferation [78]. Nucleotide triphosphates (ATP, GTP) have high potency in inhibiting NK cell-mediated cytotoxicity, this tendency however decreases with reduced negative charge due to less phosphate group [(ADP, GDP) > (AMP, GMP)]. Ectonucleotidases do not have any significant role in modulating the cytolytic effect of NK cells by extracellular ATP/ADP/AMP [79]. NK cells express lower level of CD73 even with IL-15 and IL-12 priming [74]. Decreased expression of P2Y6 promotes the development of the NK precursor cells into immature NK and mature NK cells suggesting P2Y6 as a negative regulator of NK cell maturation and function [80]. Among other extracellular purine derivatives, NKT cells display higher sensitivity to NAD+ and induce cell death via P2X7 pathway [81, 82]. Furthermore, another phenotypically heterogeneous NKT cells subset includes invariant natural killer T (iNKT), which are CD4 and CD8 negative but express NK cell marker and produce IL-4 and IFNγ. iNKT recognizes lipid antigens combined with CD1d on the surface [83]. Activation of iNKTs *in vitro* induces the expression of purinergic signaling genes A2A, P2X7R, CD38, CD39, NPP1, CD73, PANX-1, and ENT1, which has an anti-inflammatory role [84]. iNKT cells interact with DCs and monocytes via P2X7 dependent and an independent manner, respectively [85, 86]. Overall, NK cells alter their functional responses to adenosine signaling via mechanisms that are sensitive to specific cytokine activation programs.

3.2 Effect on adaptive immune signaling

3.2.1 T cells

Activation of T cell immune response is the key in adaptive immune system functions, which elicits both cellular and humoral immunity. Naïve T cells are activated by APCs, but they require two subsequent signals, first one is the binding of TCR to peptide–MHC complex and the second one is the co-stimulatory interaction at the interface between APCs and T cells via B7/CD28, LFA-1/ICAM-1 and ICAM2, and CD2/LFA-3 ligand and receptor complex [87]. Di Virgilio et al. was the first to show T cell responsiveness to extracellular ATP (eATP), back to 1989 [88]. Once T cells are activated, they release ATP via PANX-1 channels, resulting in the activation of P2X1, P2X4, and P2X7 receptors that promotes downstream signal transduction pathways leading to IL-2 expression and T cell proliferation via Ca^{+2} influx [89]. P2X7 receptor stands out among P2X family members as the most important regulator of T cell function [90]. The released ATP stimulates purinergic receptors that also contributes to the amplification of co-stimulatory TCR/CD28 signal at the immune synapse by autocrine stimulation of P2X7 [91] and P2Y1 receptors [92]. In addition, the T cell activation via P2X7R inhibits the immunosuppressive Tregs cells [93]. P2X7R is also crucial for the activation of CD8 T cells, and its expression increases as they differentiate to T_{CM} (central memory) and T_{RM} (tissue-resident memory) suggesting its key role in generating long-lived memory CD8T cells [94]. However, another study demonstrated that eATP treatment can trigger cell death in the naive CD8 ($CD44^{lo}CD45RB^{hi}$) subset, but it is unable to induce these cellular activities in the effector/memory CD8 ($CD44^{hi}CD45RB^{hi}$) subset. Even though both subsets express similarly low levels of P2X7R, but they demonstrate different sensitivity to ATP depending on the stage of differentiation instead of P2X7R expression levels [95]. Importantly, expression of CD39 and CD73, the ecto-5′-nucleotidase that degrades extracellular AMP into adenosine, by other immune and tissue-resident cells can dramatically condition the outcome of T cell responses [96]. On the other hand, A2A receptor signal inhibits Th1 cell generation and IFN-γ production, triggering the induction of FoxP3 + Treg cell subset and the production of TGF-β. ATP catabolism and generation of retaliatory metabolite adenosine is a typical suppression mechanism of regulatory cells involving Treg, type-1 regulatory (Tr1) T cells, and myeloid-derived suppressor cells (MDSCs) [97]. These regulatory cells express CD39 and CD73 to abrogate ATP-related effects and enable the inhibitory properties. P2X7R can imprint distinct outcomes to the T cell depending on the metabolic fitness and/ or developmental stage via autocrine signaling or microenvironment's clues. The peculiarity of P2X7R function as cationic channel and cytolytic pore could be responsible for some apparently contradictory findings on P2X7R dependent responses in particular T cell subsets in different experimental settings [94–96].

3.2.2 B cells

Another important arm of adaptive immune response is the humoral immunity, which is mediated by B cells. These cells are also necessary for the development of T-cell immunity because they serve as an APC, providing costimulatory signals and producing cytokines necessary for effector functions of T cells. B cells exhibit expression of the membrane B cell receptor (BCR), which can recognize antigens in their native forms, thus B cells do not need antigen presentation for activation. Antigen recognition, together with signals from activated Th2 cells, induces B cells to proliferate and generate effector plasma cells and memory B cells. B cells expresses ectonucleotidases—CD39 and CD73, P1 receptors—A1, A2A, and A3, and P2 receptors—P2X1, P2X2, P2X4, and P2X7 [33, 98, 99]. The function and activity of B cells are largely governed by the concentration of adenosine and ATP in the microenvironment. Adenosine imposes suppressive effect on B cells, whereas increased ATP release and production are associated with activated B cells thereby exerting pro-inflammatory effect on the target tissue and IgM release [100]. *In vitro* activated B cells exhibit downregulation of CD73, which mainly produces AMP, and inhibits T-cell proliferation and cytokine production, whereas overexpress A3 receptor in activated state [98]. Accumulation of pericellular ATP occurring in B cells activates the P2X7 receptor, which results in shedding of CD21, CD23, and CD62L from the cell surface [101, 102]. This process is involved in transendothelial migration of B cells. There is also evidence showing that P2X7 is directly involved in the release of IgM from B cells after T cell independent activation [103]. Moreover, CD73 is progressively upregulated on germinal center (GC) B cells following immunization, and is expressed at even higher levels among T follicular helper cells but is absent among plasma cells and plasmablasts. CD73-dependent adenosine signaling is prominent in the mature GC, maintenance of plasma cell compartment and necessary for immunoglobulin class switching [100, 104]. Thus, any disruption in the balance of ATP signaling that is dominant in activated B cells and adenosine signaling, which seems crucial in achieving immunocompetence by activated cells [100] could lead to severe immunological disorder.

4. Implication of purinergic signaling in various pathological conditions

A healthy individual has a practically insignificant amount ATP in the extracellular microenvironment (at the nanomolar range), whereas, they have significantly higher concentration of ATP in the intracellular environment (reaching several millimolar), as ATPs are the powerhouse of the cell. Inflammatory stress due to the increased production of proinflammatory mediators associates with release of ATP and other nucleotides into the extracellular space (**Figure 1**). These extracellular nucleotides trigger a stimulation of purinergic receptors, which is a normal physiological phenomenon and beneficial for preventing tissue damage ensuring host survival, it may also be detrimental for clearance of pathogens or dying cells. However, failure in the fine tuning of the immune response alters inflammatory and regulatory microenvironments, leading to unbalanced stimulation and culminates a hyperinflammatory condition generating numerous pathologies such as autoimmunity, chronic infectious diseases, and cancer (**Figure 2**).

4.1 Purinergic signaling in autoimmune disease

Autoimmune diseases are characterized by diverse clinical manifestations including dysregulated innate and adaptive immune signaling, chronic inflammation, autoreactive immune cells, generation of autoantibodies to self-nuclear and cytoplasmic component. Based on the target organ and tissues, they are represented as systemic lupus erythematosus (SLE), rheumatoid arthritis (RA), multiple sclerosis (MS), Sjogren's syndrome (SS), systemic sclerosis (SSc), etc. As described previously that some of the purinergic receptors are coupled with inflammasome assembly, pro-inflammatory cascades, secretion of IL-1β, IL-18, and T and B cell activation, and maturation, all these events play a pivotal role in autoimmunity [105].

4.1.1 Systemic lupus erythematosus

SLE is an inflammatory autoimmune disease that affects many organs, including the skin, joints, the central nervous system, and the kidneys. A frequent and serious manifestation of SLE includes glomerulonephritis (GN), a condition that can cause proteinuria and progresses to kidney failure. These diverse clinical features include hematological and serological abnormalities, such as decreased levels of complement and increased levels of autoantibodies [106, 107]. SLE has multiple etiology like genetic, environmental, and hormonal factor but involvement of dysfunctional innate and the adaptive system is prominent [108]. Purinergic signaling is another key pathway that connects with the inflammatory signaling cascade and contributes to the immunopathogenesis of SLE.

Till date, more than 180 autoantibodies have been documented in SLE patients [107]. Source of the diverse pool of autoantigens are apoptosis [109], netosis [110], and pyroptosis [111]. Simultaneously, SLE patient also exhibit impairment of phagocytotic clearance and NET degradation [112, 113], which together represent a mechanism that trigger to breakdown of the self-tolerance against autoantigens and leading to initiation of SLE. Defects in the purinergic signaling and its role in SLE pathogenesis and disease severity had been described at several instances. Therefore, P2X7R activation by ATP or by extracellular complexes, such as NETs, might have a dual pathogenetic role in promoting inflammation in lupus: on one hand, it directly triggers inflammation by stimulating the NLRP3 inflammasome, and on the other it has an indirect pro-inflammatory effect by inducing pyroptotic cell death [114]. Presence of the NETs in the microenvironment induce NLRP3 inflammasome, in macrophages and results in the amplification of inflammation by releasing of IL-1β and IL-18, which is mediated via P2X7R [115]. Induction of inflammasome and IL-1β and IL-18 release have been shown to contribute to the cardiovascular, skin, and nephritis manifestations [116–118]. Evidence suggests the higher P2X7R in renal tissue of lupus nephritis patients [119]. In that context, a study demonstrated substantial up-regulation of P2X7R, NLRP3, and ASC, in the kidneys of MLR/lpr mice compared to control mice and inhibition of P2X7R ameliorates the disease phenotype mainly diminished both the severity of nephritis and levels of circulating anti-dsDNA antibodies [120, 121]. The presence of single nuclear polymorphism (SNP) 489C>T in P2X7 receptor had been associated with increased inflammasome activation in SLE patients and shows involvement in pericarditis [122, 123]. Th1, Th17, and Regulatory T (Treg) cells in SLE patients display higher expression of P2X7 receptor, which correlates with active SLE disease and increased levels of IFN-γ, IL-1β, IL-6, IL-17A, and IL-23 cytokines [124]. Monocytes and lymphocytes from SLE patients and RA patients show reduced expression of P2X7R gene. They show reduced tendency to induce apoptosis and cytokine release *in vitro* compared to cells from healthy individual [125].

Furthermore, P2X7R has an important function of restricting the expansion of T follicular helper (Tfh) cells by pyroptosis and controls the development of pathogenic ICOS+ IFN-γ–secreting cells and in turn prevents the overproduction of autoantibodies and activation of T cells that ultimately controls the production of autoantibodies conditions [126]. SLE patients exhibit deletion of P2X7R genes that have deleterious effect of autoantibody generation [126]. Another study had reported that deletion of P2X7R could amplify the defect in peripheral T cell homeostasis due to the FAS mutation and thus contribute to the autoimmune pathology [127]. In similar way, another purinergic receptor P2Y8R restricts the proliferation of self-tolerant B cells. Distinct variant of P2Y8R had been shown to be downregulated in SLE patients and these are associated with the loss of function, which leads to increased expansion of self-reactive B cells, resulting in the increased autoantibody production. P2Y8R correlated with lupus nephritis and increased age-associated B cells and plasma cells indicating a role of P2Y8R in immunological tolerance and lupus pathogenesis [128].

The role of CD39 in the maintenance of immune tolerance is associated with its capacity of degrading ATP and consequently inhibiting the production of IL-17, which stimulates B cells to produce autoantibodies. Ectonucleotide provides protection in by converting eATP to adenosine. Deletion of ectonucleotides mainly, CD39 and CD73 lead to higher levels of anti-RNP antibodies in response to pristane, with

CD73 deletion in particular promoting expansion of splenic B cell and T cell populations that likely contribute to autoantibody production [129]. B cells show the highest CD73 surface expression among human circulating immune cells. In SLE patients, the activity of CD73 and CD38 was found to be selectively silenced in B cells. Since CD73 is the bottleneck of extracellular nucleotide degradation to anti-inflammatory adenosine, this pathway is likely to be a crucial step in the pathophysiology of SLE involving B cell immune cell interactions [130].

4.1.2 Rheumatoid arthritis

RA is a chronic inflammatory disease of joints characterized by damage of bone and cartilage, which leads to joint destruction and disability. Primarily, it is driven by proliferation of synovial fibroblasts, inflammatory response of innate and adaptive immune response, differentiation of macrophage into osteoclasts, and impaired differentiation of mesenchymal stem cells into osteoblasts. The incidence is about 5 per 1000 people and can lead to severe joint damage and disability [131]. Studies have shown a critical role for P2 receptors in osteoblastogenesis and mineralization, synoviocytes proliferation, inflammation of immune cells, and differentiation of macrophages into osteoclasts [132]. Specifically, P2X7, P2Y14, P2Y12, P2Y6, P2Y1, P2Y2, and P2X4 receptors are involved in modulating bone and joint biology [133]. Pain is the major symptom of RA, which associates with the involvement of P2X4R had been reported in chronic arthritis [134]. Knockout of this gene in mice model alleviates the pain [135]. P2X4R control the production of Th17 cells, as shown by the inhibition of P2X4 receptor which reduced the production of IL-17 but not of IFN-γ by effector/memory CD4+ T cells isolated from patients with rheumatoid arthritis [136]. Inhibition of P2X4R associated with the attenuation of synovial inflammation and joint destruction as well as decreased the levels of serum IL-1β, TNF-α, IL-6, and IL-17 via NLRP1 [137]. Similar to SLE, SNP in P2X7 is associated with increased inflammatory response and susceptibility to RA [123, 125]. P2X7 receptor-mediates the release of cathepsins from macrophages is a cytokine-independent mechanism potentially involved in joint diseases and is important for osteoclastogenesis [138]. It also regulates the differentiation of Th17 cells and type II collagen-induced arthritis in mice [139]. P2Y receptor also contribute to the development of RA such as P2Y11 receptor induce inflammation in primary fibroblast-like synoviocytes [140], P2Y12 and P2Y14 receptors induce bone lysis by activating osteoclasts [141–143]. RA patients demonstrate differential expression of adenosine receptors on synovium with preferential expression of A3 and its variant. However, in a separate RA cohort treated with methotrexate shows overexpression A2A and A2B indicating the anti-inflammatory property via these adenosine receptors [144]. Under hypoxia condition, bone resorption is increased in RA patients via A2B receptors. Inhibition of A2B receptors potentially prevent the hypoxia-mediated pathological osteolysis in RA [145].

The expression of CD39 in Tregs is limited by single nucleotide polymorphisms (SNP). It has been shown that AA genotype of the rs10748643 SNP, a low-expressing CD39 variant, is involved in the regulation of the immune system in autoimmunity [146]. A reduced response to methotrexate (MTX) in patients with rheumatoid arthritis was also shown to be related to an SNP that decreases the frequencies of CD39-expressing Tregs, the rs7071836 SNP [147]. Lower expression of CD73 in lymphocytes at the sites of inflammation has been associated with disease severity in juvenile idiopathic arthritis [148].

4.1.3 Multiple sclerosis

Multiple sclerosis (MS) is a chronic inflammatory demyelinating disease of the central nervous system, characterized by the presence of focal lesions in white and gray matter, which is associated with pathological and progression neurological dysfunction. Presence of peripheral immune cells infiltration is a main diagnostic hallmark of the disease. Purinergic receptors control immune cell function as well as neuronal and oligodendroglia survival, and the activation of astrocytes and microglia, the endogenous brain immune cells. Genetic variation in P2X4 and P2X7 receptors show susceptibility to MS. Functionally, the variants impair the expression of P2X7 on the surface resulting in the inhibition of ATP-induced pore function and phagocytic activity [149]. Cortical microglia from MS patient exhibit loss of P2Y12 receptor, which associates with the pro-inflammatory and neuronal damaging profile in MS [150]. On the other hand, P2Y12 is the markers of platelet and megakaryocyte activation. Its increased expression in MS patients associates with cardiovascular disease [151]. A study in mice model show that the loss of P2Y6 develop more severe experimental autoimmune encephalomyelitis compared with wild-type mice as it has pivotal role in DCs regulation [64]. Lymphocytes from MS patients also exhibit upregulation of A2A receptor, which modulates the release of proinflammatory cytokine TNF-α, IFN-γ, IL-6, IL-1β, IL-17 via NF-κB. A2A receptor upregulation was observed in lymphocytes from MS patients in comparison with healthy subjects. The stimulation of these receptors mediated a significant inhibition of TNF-α, IFN-γ, IL-6, IL-1β, IL-17, and cell proliferation as well as very late antigen (VLA)-4 expression and NF-κB activation [152].

CD39 expressing Treg cells controls the neuroinflammation in MS by suppressing the pathogenic Th17 cells and IL-17 production [31]. Its activity and the frequency were elevated in relapsing MS patients [153]. Furthermore, a study on animal model demonstrated that overexpression of CD39 on reactive microglia/macrophages that associates with either pro-inflammatory (M1-subtype) or neuroprotective (M2-subtype) at different stages of the disease. At the peak of EAE, CD39 immunoreactivity showed much higher co-occurrence with Arg1 immunoreactivity in microglia and macrophages, compared to iNOS, implying its stronger association with M2-like reactive phenotype [154]. Thus, modulation of purinergic signaling using an agonist or antagonist provides a new avenue for treatment of disease [155, 156].

4.2 Purinergic signaling during bacterial and viral infection

Infectious diseases are caused by the invasion of pathogenic microorganisms. After infection, host immune system elicits the anti-microbial immune response and at the same time microorganisms develop strategies to evade host defense mechanism. This involves generation of a variety of inflammatory and suppressive responses along with regulatory feedback systems to eliminate the pathogens but also to restore the homeostatic condition following infection or injury [157]. The purinergic system has the dual function of regulating the immune response and triggering effector antimicrobial response against bacterial and viral infections. During the infections, the ATP release initiates a cascade that activates purinergic receptors. This receptor activation enhances the secretion of pro-inflammatory cytokines and performs the chemotaxis of macrophages and neutrophils, generating an association between the immune and the purinergic systems. Immunomodulation by purinergic signaling has been widely discussed elsewhere [26, 158]. Some instances of involvement of purinergic signaling in bacterial infection include, reduced CD73 expression was associated with macrophage phagocytosis and an efficient clearance of *Salmonella* infection [159]. Likewise, depletion of CD39 on CD4, CD8, and Treg cells augments the T cells response to *Listeria* and *Mycobacterium* infections [160, 161]. On the other hand, transgenic mice with overexpression of CD39 in lung epithelia shows increased recruitment of neutrophils and macrophages in lungs upon *Pseudomonas aeruginosa* infection. The CD39 activity associates with efficient clearance of infection [162]. CD39, due to ATP-scavenging property it limits P2X7 receptor mediated pro-inflammatory responses. Thus, deletion of CD39 exacerbates sepsis-induced liver injury [163]. P2X7R signaling has a detrimental role in severe tuberculosis infection. ATP release and activation of P2X7R cause macrophage necrosis resulting in the spread of bacterial particles, leukocyte infiltration, and tissue damage [164]. Deletion of P2X7 receptor or blockage of P2X7R, or scavenging of eATP may attenuated inflammation, largely preventing increased cytokine secretion and tissue damage [163, 165].

Immunomodulation of purinergic signaling had been implicated in wide variety of viral infections such as human immunodeficiency virus (HIV)-1, hepatitis virus, dengue virus, and SARS-CoV2 [166–169]. HIV-1 primarily infects CD4 T cells, but also affects myeloid dendritic cells and monocyte, macrophages populations that express CD4 receptor. Infected patients exhibit decreased CD4 T cell counts and a reversed CD4/CD8 T cells ratio. Adenosine has an immunosuppressive effect, patients with HIV infection show upregulated CD39 on Treg cells which is inversely related with the CD4 T cell count [167, 170]. In contrast to CD39, CD73 expression was diminished on CD4 T cells, which represent a phenotypically and functionally different subpopulation of CD73+ CD4 T cells. This T cell subsets are preferentially reduced in HIV patients, which suggests the effect of an adenosine diminished microenvironment that cannot prevent persistent immune activation. CD73+ CD4+ T cell counts were inversely associated with T cell activation, as well as plasma C reactive protein levels [171]. Besides, CD73 is involved in the expansion of HIV-specific CD8 T cells, whereas CD73 expression is higher in memory CD8+ T cell subset. The frequency of CD73+ CD8+ T cells is inversely associated with cell activation and plasma viral load [172]. PANX-1 hemichannel opening, activation of P2Y2R, P2X1R are involved in the mediating the effective viral entry and replication in CD4 or target cells [173–175]. Blocking the P2X1 and P2X7 receptors inhibits the viral entry and fusion [176]. Similarly, the P2X1R, P2X4R and P2X7R expression increased in during hepatitis C virus infection and Dengue virus infection [168, 177]. Blocking P2X receptor with antagonist improves the anti-viral response and T cell function [168, 178].

Given the pathophysiological role of purinergic signaling in highly prevalent viral infections has developed a potential interest in investigating the effects of purinergic system in severe acute respiratory syndrome coronavirus 2 virus (SARS-CoV-2). SARS-CoV-2 infection had impacted more than millions of people worldwide since its emergence in December 2019, in Wuhan, China. The clinical manifestations of SARS-CoV-2 include pneumonia, acute respiratory distress syndrome (ARDS), and hyperinflammation. SARS-CoV-2 primarily invade the alveolar epithelia of respiratory tract and lungs where they replicate, triggers the activation of the immune system resulting in the release of cytokines as a defense mechanism, but the response become exaggerated and prompt the so-called "cytokine storm." This is a state of hyperinflammatory response, which develops acute respiratory syndrome (SARS). This is characterized by fever, cough, and difficulty breathing, which can progress to pneumonia, failure of different organs, and death. Patients with SARS-CoV-2 infection exhibit increased purinergic signaling, which has been suggested to have a role in hyperinflammatory state [179]. The mechanisms have been described in very detail in review articles [169, 180]. The increased inflammations resulting from activated purinergic signaling in SARS-Cov-2 infections are also associated with different pathological conditions such as neuropathy [181], thrombopathy [182, 183]. As observed in other viral infections patients with SARS-CoV-2 shows reduced expression of CD73 on circulating CD8, NK, and NKT cells. However, cells lacking CD73 exhibit increased cytotoxic effector capacity compared to their counterpart CD73+ [184]. P2X7R-NLRP3 signaling axis are the key driver of inflammation in SARS-CoV-2 [185]. Therefore, P2X7R could serve as a potential therapeutic target to control the inflammation [186]. The readily available and affordable P2X7R antagonist lidocaine can abrogate hyperinflammation and restore the normal immune function [169]. Understanding this biology is very crucial as anti-inflammatory drugs are not effective and sometimes accompanied by serious adverse effects.

5. Conclusion

This chapter has highlighted the importance of purinergic signaling in modulating the immune system in various therapeutic areas. Purinergic system is capable of fine tuning the levels of nucleotides and their derivative in the extracellular space thereby controlling the chemotaxis, proliferation, differentiation of various immune cell presents locally or far from the infectious site. Dysregulation of purinergic signaling because of genetic factor or escape mechanism employed by the microbes or regulatory cell leads to overt inflammation that contributes to the disease. Special attention has been paid to the mechanisms through which alterations in the various compartments of the purinergic system could contribute to the patho-pathophysiology of autoimmune disease and microbial infection. This chapter could help in gaining insight on the possibility of counteracting such dysfunctions by means of pharmacological interventions on purinergic molecular targets.

Author details

Richa Rai
Division of Hematology and Oncology, Department of Medicine, Tisch Cancer Institute, Icahn School of Medicine at Mount Sinai, New York, NY, USA

*Address all correspondence to: richa.23m@gmail.com

References

[1] Dunn J, Grider MH. Physiology, adenosine triphosphate. In: StatPearls [Internet]. Treasure Island (FL): StatPearls Publishing; 2021

[2] Drury AN, Szent-Györgyi A. The physiological activity of adenine compounds with especial reference to their action upon the mammalian heart. The Journal of Physiology. 1929;**68**(3):213-237

[3] Burnstock G. Purinergic nerves. Pharmacological Reviews. 1972;**24**(3):509-581

[4] Burnstock G. The past, present and future of purine nucleotides as signalling molecules. Neuropharmacology. 1997;**36**(9):1127-1139

[5] Burnstock G. Purinergic system. In: Offermanns S, Rosenthal W, editors. Encyclopedia of Molecular Pharmacology. Berlin, Heidelberg, Springer; 2008. pp. 1047-1053

[6] Burnstock G. Short- and long-term (trophic) purinergic signalling. Philosophical Transactions of the Royal Society of London. Series B, Biological Sciences. 2016;**371**(1700):20150422

[7] Burnstock G. The therapeutic potential of purinergic signalling. Biochemical Pharmacology. 2018;**151**:157-165

[8] Burnstock G, Verkhratsky A. Long-term (trophic) purinergic signalling: Purinoceptors control cell proliferation, differentiation and death. Cell Death & Disease. 2010;1(1):e9-e

[9] Huang Z, Xie N, Illes P, Di Virgilio F, Ulrich H, Semyanov A, et al. From purines to purinergic signalling: Molecular functions and human diseases. Signal Transduction and Targeted Therapy. 2021;**6**(1):162

[10] Fitz JG. Regulation of cellular ATP release. Transactions of the American Clinical and Climatological Association. 2007;**118**:199-208

[11] Dosch M, Gerber J, Jebbawi F, Beldi G. Mechanisms of ATP release by inflammatory cells. International Journal of Molecular Sciences. 2018;**19**(4):1222

[12] Taruno A. ATP release channels. International Journal of Molecular Sciences. 2018;**19**(3):808

[13] Imura Y, Morizawa Y, Komatsu R, Shibata K, Shinozaki Y, Kasai H, et al. Microglia release ATP by exocytosis. Glia. 2013;**61**(8):1320-1330

[14] Südhof TC, Rothman JE. Membrane fusion: Grappling with SNARE and SM proteins. Science. 2009;**323**(5913):474-477

[15] Wang N, De Bock M, Decrock E, Bol M, Gadicherla A, Vinken M, et al. Paracrine signaling through plasma membrane hemichannels. Biochimica et Biophysica Acta. 2013;**1828**(1):35-50

[16] Kar R, Batra N, Riquelme MA, Jiang JX. Biological role of connexin intercellular channels and hemichannels. Archives of Biochemistry and Biophysics. 2012;**524**(1):2-15

[17] Wang N, De Bock M, Decrock E, Bol M, Gadicherla A, Bultynck G, et al. Connexin targeting peptides as inhibitors of voltage- and intracellular Ca^{+2}-triggered Cx43 hemichannel opening. Neuropharmacology. 2013;**75**:506-516

[18] Dourado M, Wong E, Hackos DH. Pannexin-1 is blocked by its C-terminus through a delocalized non-specific interaction surface. PLoS One. 2014;**9**(6): e99596

[19] Penuela S, Gehi R, Laird DW. The biochemistry and function of pannexin channels. Biochimica et Biophysica Acta. 2013;**1828**(1):15-22

[20] Chekeni FB, Elliott MR, Sandilos JK, Walk SF, Kinchen JM, Lazarowski ER, et al. Pannexin 1 channels mediate 'find-me' signal release and membrane permeability during apoptosis. Nature. 2010;**467**(7317):863-867

[21] Sandilos JK, Chiu YH, Chekeni FB, Armstrong AJ, Walk SF, Ravichandran KS, et al. Pannexin 1, an ATP release channel, is activated by caspase cleavage of its pore-associated C-terminal autoinhibitory region. The Journal of Biological Chemistry. 2012;**287**(14):11303-11311

[22] Yang D, He Y, Muñoz-Planillo R, Liu Q, Núñez G. Caspase-11 requires the Pannexin-1 channel and the purinergic P2X7 pore to mediate pyroptosis and endotoxic shock. Immunity. 2015;**43**(5):923-932

[23] Iyer SS, Pulskens WP, Sadler JJ, Butter LM, Teske GJ, Ulland TK, et al. Necrotic cells trigger a sterile inflammatory response through the Nlrp3 inflammasome. Proceedings of the National Academy of Sciences of the United States of America. 2009;**106**(48):20388-20393

[24] Elliott MR, Chekeni FB, Trampont PC, Lazarowski ER, Kadl A, Walk SF, et al. Nucleotides released by apoptotic cells act as a find-me signal to promote phagocytic clearance. Nature. 2009;**461**(7261):282-286

[25] Stefan C, Jansen S, Bollen M. Modulation of purinergic signaling by NPP-type ectophosphodiesterases. Purinergic Signalling. 2006;**2**(2):361-370

[26] Giuliani AL, Sarti AC, Di Virgilio F. Ectonucleotidases in acute and chronic inflammation. Frontiers in Pharmacology. 2021;**11**:619458

[27] Haas CB, Lovászi M, Pacher P, de Souza PO, Pelletier J, Leite RO, et al. Extracellular ectonucleotidases are differentially regulated in murine tissues and human polymorphonuclear leukocytes during sepsis and inflammation. Purinergic Signalling. 2021;**17**(4):713-724

[28] Haas CB, Lovászi M, Braganhol E, Pacher P, Haskó G. Ectonucleotidases in inflammation, immunity, and cancer. Journal of Immunology. 2021;**206**(9):1983-1990

[29] Borsellino G, Kleinewietfeld M, Di Mitri D, Sternjak A, Diamantini A, Giometto R, et al. Expression of ectonucleotidase CD39 by Foxp3+ Treg cells: Hydrolysis of extracellular ATP and immune suppression. Blood. 2007;**110**(4):1225-1232

[30] Mandapathil M, Hilldorfer B, Szczepanski MJ, Czystowska M, Szajnik M, Ren J, et al. Generation and accumulation of immunosuppressive adenosine by human CD4+ CD25highFOXP3+ regulatory T cells. The Journal of Biological Chemistry. 2010;**285**(10):7176-7186

[31] Fletcher JM, Lonergan R, Costelloe L, Kinsella K, Moran B, O'Farrelly C, et al. CD39+Foxp3+ regulatory T cells suppress pathogenic Th17 cells and are impaired in multiple sclerosis. Journal of Immunology. 2009;**183**(11):7602-7610

[32] Schneider E, Winzer R, Rissiek A, Ricklefs I, Meyer-Schwesinger C, Ricklefs FL, et al. CD73-mediated

adenosine production by CD8 T cell-derived extracellular vesicles constitutes an intrinsic mechanism of immune suppression. Nature Communications. 2021;**12**(1):5911

[33] Saze Z, Schuler PJ, Hong C-S, Cheng D, Jackson EK, Whiteside TL. Adenosine production by human B cells and B cell-mediated suppression of activated T cells. Blood. 2013;**122**(1):9-18

[34] Ferrari D, Pizzirani C, Adinolfi E, Lemoli RM, Curti A, Idzko M, et al. The P2X7 receptor: A key player in IL-1 processing and release. Journal of Immunology. 2006;**176**(7):3877-3883

[35] Clavarino G, Delouche N, Vettier C, Laurin D, Pernollet M, Raskovalova T, et al. Novel strategy for phenotypic characterization of human B lymphocytes from precursors to effector cells by flow cytometry. PLoS One. 2016;**11**(9):e0162209

[36] Sandoval-Montes C, Santos-Argumedo L. CD38 is expressed selectively during the activation of a subset of mature T cells with reduced proliferation but improved potential to produce cytokines. Journal of Leukocyte Biology. 2005;**77**(4):513-521

[37] Piedra-Quintero ZL, Wilson Z, Nava P, Guerau-de-Arellano M. CD38: An immunomodulatory molecule in inflammation and autoimmunity. Frontiers in Immunology. 2020;**11**:597959

[38] Partida-Sánchez S, Cockayne DA, Monard S, Jacobson EL, Oppenheimer N, Garvy B, et al. Cyclic ADP-ribose production by CD38 regulates intracellular calcium release, extracellular calcium influx and chemotaxis in neutrophils and is required for bacterial clearance in vivo. Nature Medicine. 2001;**7**(11):1209-1216

[39] Kang J, Park KH, Kim JJ, Jo EK, Han MK, Kim UH. The role of CD38 in Fcγ receptor (FcγR)-mediated phagocytosis in murine macrophages. The Journal of Biological Chemistry. 2012;**287**(18):14502-14514

[40] Muñoz P, Mittelbrunn M, de la Fuente H, Pérez-Martínez M, García-Pérez A, Ariza-Veguillas A, et al. Antigen-induced clustering of surface CD38 and recruitment of intracellular CD38 to the immunologic synapse. Blood. 2008;**111**(7):3653-3664

[41] Linden J, Koch-Nolte F, Dahl G. Purine release, metabolism, and signaling in the inflammatory response. Annual Review of Immunology. 2019;**37**:325-347

[42] Zeidler JD, Hogan KA, Agorrody G, Peclat TR, Kashyap S, Kanamori KS, et al. The CD38 glycohydrolase and the NAD sink: Implications for pathological conditions. American Journal of Physiology. Cell Physiology. 2022;**322**(3):C521-C545

[43] Albright RA, Ornstein DL, Cao W, Chang WC, Robert D, Tehan M, et al. Molecular basis of purinergic signal metabolism by ectonucleotide pyrophosphatase/phosphodiesterases 4 and 1 and implications in stroke. The Journal of Biological Chemistry. 2014;**289**(6):3294-3306

[44] Knowlden S, Georas SN. The autotaxin-LPA axis emerges as a novel regulator of lymphocyte homing and inflammation. Journal of Immunology (Baltimore, MD: 1950). 2014;**192**(3):851-857

[45] Benesch MG, Tang X, Dewald J, Dong WF, Mackey JR, Hemmings DG, et al. Tumor-induced inflammation in mammary adipose tissue stimulates a vicious cycle of autotaxin expression and breast cancer progression. The FASEB Journal. 2015;**29**(9):3990-4000

[46] Burnstock G. Purine and purinergic receptors. Brain and Neuroscience Advances. 2018;**2**:2398212818817494

[47] Pastor-Anglada M, Pérez-Torras S. Who is who in adenosine transport. Frontiers in Pharmacology. 2018;**9**:627

[48] Antonioli L, Colucci R, Pellegrini C, Giustarini G, Tuccori M, Blandizzi C, et al. The role of purinergic pathways in the pathophysiology of gut diseases: Pharmacological modulation and potential therapeutic applications. Pharmacology & Therapeutics. 2013;**139**(2):157-188

[49] Marshall JS, Warrington R, Watson W, Kim HL. An introduction to immunology and immunopathology. Allergy, Asthma & Clinical Immunology. 2018;**14**(2):49

[50] Cekic C, Linden J. Purinergic regulation of the immune system. Nature Reviews Immunology. 2016;**16**(3):177-192

[51] Kronlage M, Song J, Sorokin L, Isfort K, Schwerdtle T, Leipziger J, et al. Autocrine purinergic receptor signaling is essential for macrophage chemotaxis. Science Signaling. 2010;**3**(132):ra55

[52] Chen Y, Corriden R, Inoue Y, Yip L, Hashiguchi N, Zinkernagel A, et al. ATP release guides neutrophil chemotaxis via P2Y2 and A3 receptors. Science. 2006;**314**(5806):1792-1795

[53] Cohen HB, Briggs KT, Marino JP, Ravid K, Robson SC, Mosser DM. TLR stimulation initiates a CD39-based autoregulatory mechanism that limits macrophage inflammatory responses. Blood. 2013;**122**(11):1935-1945

[54] Ferrante CJ, Pinhal-Enfield G, Elson G, Cronstein BN, Hasko G, Outram S, et al. The adenosine-dependent angiogenic switch of macrophages to an M2-like phenotype is independent of interleukin-4 receptor alpha (IL-4Rα) signaling. Inflammation. 2013;**36**(4):921-931

[55] Wang L, Jacobsen SE, Bengtsson A, Erlinge D. P2 receptor mRNA expression profiles in human lymphocytes, monocytes and CD34+ stem and progenitor cells. BMC Immunology. 2004;**5**:16

[56] Klaver D, Thurnher M. Control of macrophage inflammation by P2Y purinergic receptors. Cell. 2021;**10**(5):109

[57] Merz J, Nettesheim A, von Garlen S, Albrecht P, Saller BS, Engelmann J, et al. Pro- and anti-inflammatory macrophages express a sub-type specific purinergic receptor profile. Purinergic Signalling. 2021;**17**(3):481-492

[58] Schnurr M, Toy T, Shin A, Hartmann G, Rothenfusser S, Soellner J, et al. Role of adenosine receptors in regulating chemotaxis and cytokine production of plasmacytoid dendritic cells. Blood. 2004;**103**(4):1391-1397

[59] Panther E, Corinti S, Idzko M, Herouy Y, Napp M, la Sala A, et al. Adenosine affects expression of membrane molecules, cytokine and chemokine release, and the T-cell stimulatory capacity of human dendritic cells. Blood. 2003;**101**(10):3985-3990

[60] Wilson JM, Kurtz CC, Black SG, Ross WG, Alam MS, Linden J, et al. The A2B adenosine receptor promotes Th17 differentiation via stimulation of dendritic cell IL-6. Journal of Immunology. 2011;**186**(12):6746-6752

[61] Ring S, Pushkarevskaya A, Schild H, Probst HC, Jendrossek V, Wirsdörfer F, et al. Regulatory T cell-derived adenosine induces dendritic cell migration through

the Epac-Rap1 pathway. Journal of Immunology. 2015;**194**(8):3735-3744

[62] Sáez PJ, Vargas P, Shoji KF, Harcha PA, Lennon-Duménil AM, Sáez JC. ATP promotes the fast migration of dendritic cells through the activity of pannexin 1 channels and P2X(7) receptors. Science Signaling. 2017;**10**(506):eaah7107

[63] Yu Y, Feng S, Wei S, Zhong Y, Yi G, Chen H, et al. Extracellular ATP activates P2X7R-NF-κB (p65) pathway to promote the maturation of bone marrow-derived dendritic cells of mice. Cytokine. 2019;**119**:175-181

[64] Li Z, He C, Zhang J, Zhang H, Wei H, Wu S, et al. P2Y(6) deficiency enhances dendritic cell-mediated Th1/Th17 differentiation and aggravates experimental autoimmune encephalomyelitis. Journal of Immunology. 2020;**205**(2):387-397

[65] Li R, Wang J, Li R, Zhu F, Xu W, Zha G, et al. ATP/P2X7-NLRP3 axis of dendritic cells participates in the regulation of airway inflammation and hyper-responsiveness in asthma by mediating HMGB1 expression and secretion. Experimental Cell Research. 2018;**366**(1):1-15

[66] Sakaki H, Fujiwaki T, Tsukimoto M, Kawano A, Harada H, Kojima S. P2X4 receptor regulates P2X7 receptor-dependent IL-1β and IL-18 release in mouse bone marrow-derived dendritic cells. Biochemical and Biophysical Research Communications. 2013;**432**(3):406-411

[67] Silva-Vilches C, Ring S, Mahnke K. ATP and its metabolite adenosine as regulators of dendritic cell activity. Frontiers in Immunology. 2018;**9**:2581

[68] Zhao R, Qiao J, Zhang X, Zhao Y, Meng X, Sun D, et al. Toll-like receptor-mediated activation of CD39 internalization in BMDCs leads to extracellular ATP accumulation and facilitates P2X7 receptor activation. Frontiers in Immunology. 2019;**10**:2524

[69] Mantovani A, Cassatella MA, Costantini C, Jaillon S. Neutrophils in the activation and regulation of innate and adaptive immunity. Nature Reviews. Immunology. 2011;**11**(8):519-531

[70] Amulic B, Cazalet C, Hayes GL, Metzler KD, Zychlinsky A. Neutrophil function: From mechanisms to disease. Annual Review of Immunology. 2012;**30**:459-489

[71] la Sala A, Ferrari D, Di Virgilio F, Idzko M, Norgauer J, Girolomoni G. Alerting and tuning the immune response by extracellular nucleotides. Journal of Leukocyte Biology. 2003;**73**(3):339-343

[72] Chen Y, Yao Y, Sumi Y, Li A, To UK, Elkhal A, et al. Purinergic signaling: A fundamental mechanism in neutrophil activation. Science Signaling. 2010;**3**(125):ra45

[73] Kukulski F, Ben Yebdri F, Lecka J, Kauffenstein G, Lévesque SA, Martín-Satué M, et al. Extracellular ATP and P2 receptors are required for IL-8 to induce neutrophil migration. Cytokine. 2009;**46**(2):166-170

[74] Kukulski F, Bahrami F, Ben Yebdri F, Lecka J, Martín-Satué M, Lévesque SA, et al. NTPDase1 controls IL-8 production by human neutrophils. Journal of Immunology. 2011;**187**(2):644-653

[75] Pierce S, Geanes ES, Bradley T. Targeting natural killer cells for improved immunity and control of the adaptive immune response. Frontiers in cellular and infection. Microbiology. 2020;**10**:231

[76] Chambers AM, Wang J, Lupo KB, Yu H, Atallah Lanman NM, Matosevic S.

Adenosinergic Signaling alters natural killer cell functional responses. Frontiers in Immunology. 2018;**9**:2533

[77] Hoskin DW, Mader JS, Furlong SJ, Conrad DM, Blay J. Inhibition of T cell and natural killer cell function by adenosine and its contribution to immune evasion by tumor cells (review). International Journal of Oncology. 2008;**32**(3):527-535

[78] Miller JS, Cervenka T, Lund J, Okazaki IJ, Moss J. Purine metabolites suppress proliferation of human NK cells through a lineage-specific purine receptor. The Journal of Immunology. 1999;**162**(12):7376-7382

[79] Bajpai A, Brahmi Z. Regulation of resting and IL-2-activated human cytotoxic lymphocytes by exogenous nucleotides: Role of IL-2 and ecto-ATPases. Cellular Immunology. 1993;**148**(1):130-143

[80] Li Z, Gao Y, He C, Wei H, Zhang J, Zhang H, et al. Purinergic receptor P2Y(6) is a negative regulator of NK cell maturation and function. Journal of Immunology. 2021;**207**(6):1555-1565

[81] Rissiek B, Danquah W, Haag F, Koch-Nolte F. Technical advance: A new cell preparation strategy that greatly improves the yield of vital and functional Tregs and NKT cells. Journal of Leukocyte Biology. 2014;**95**(3):543-549

[82] Rissiek B, Haag F, Boyer O, Koch-Nolte F, Adriouch S. ADP-ribosylation of P2X7: A matter of life and death for regulatory T cells and natural killer T cells. In: Koch-Nolte F, editor. Endogenous ADP-Ribosylation. Cham: Springer International Publishing; 2015. pp. 107-126

[83] Krovi SH, Gapin L. Invariant natural killer T cell subsets—More than just developmental intermediates. Frontiers in Immunology. 2018;**9**:1393

[84] Yu JC, Lin G, Field JJ, Linden J. Induction of antiinflammatory purinergic signaling in activated human iNKT cells. JCI Insight. 2018;**3**(17):e91954

[85] Felley LE, Sharma A, Theisen E, Romero-Masters JC, Sauer JD, Gumperz JE. Human invariant NKT cells induce IL-1β secretion by peripheral blood monocytes via a P2X7-independent pathway. Journal of Immunology. 2016;**197**(6):2455-2464

[86] Xu X, Pocock GM, Sharma A, Peery SL, Fites JS, Felley L, et al. Human iNKT cells promote protective inflammation by inducing oscillating purinergic signaling in monocyte-derived DCs. Cell Reports. 2016;**16**(12): 3273-3285

[87] Tai Y, Wang Q, Korner H, Zhang L, Wei W. Molecular mechanisms of T cells activation by dendritic cells in autoimmune diseases. Frontiers in Pharmacology. 2018;**9**:642

[88] Di Virgilio F, Bronte V, Collavo D, Zanovello P. Responses of mouse lymphocytes to extracellular adenosine 5′-triphosphate (ATP). Lymphocytes with cytotoxic activity are resistant to the permeabilizing effects of ATP. Journal of Immunology. 1989;**143**(6):1955-1960

[89] Woehrle T, Yip L, Elkhal A, Sumi Y, Chen Y, Yao Y, et al. Pannexin-1 hemichannel-mediated ATP release together with P2X1 and P2X4 receptors regulate T-cell activation at the immune synapse. Blood. 2010;**116**(18): 3475-3484

[90] Grassi F. The P2X7 receptor as regulator of T cell development and function. Frontiers in Immunology. 2020;**11**:1179

[91] Yip L, Woehrle T, Corriden R, Hirsh M, Chen Y, Inoue Y, et al. Autocrine regulation of T-cell activation by ATP release and P2X7 receptors. The FASEB Journal. 2009;**23**(6):1685-1693

[92] Woehrle T, Ledderose C, Rink J, Slubowski C, Junger WG. Autocrine stimulation of P2Y1 receptors is part of the purinergic signaling mechanism that regulates T cell activation. Purinergic Signalling. 2019;**15**(2):127-137

[93] Schenk U, Frascoli M, Proietti M, Geffers R, Traggiai E, Buer J, et al. ATP inhibits the generation and function of regulatory T cells through the activation of purinergic P2X receptors. Science Signaling. 2011;**4**(162):ra12

[94] Borges da Silva H, Beura LK, Wang H, Hanse EA, Gore R, Scott MC, et al. The purinergic receptor P2RX7 directs metabolic fitness of long-lived memory CD8(+) T cells. Nature. 2018;**559**(7713):264-268

[95] Mellouk A, Bobé P. CD8(+), but not CD4(+) effector/memory T cells, express the CD44(high)CD45RB(high) phenotype with aging, which displays reduced expression levels of P2X(7) receptor and ATP-induced cellular responses. The FASEB Journal. 2019;**33**(3):3225-3236

[96] Safya H, Mellouk A, Legrand J, Le Gall SM, Benbijja M, Kanellopoulos-Langevin C, et al. Variations in cellular responses of mouse T cells to Adenosine-5′-triphosphate stimulation do not depend on P2X7 receptor expression levels but on their activation and differentiation stage. Frontiers in immunology. 2018;**9**:360

[97] Ohta A, Ohta A, Madasu M, Kini R, Subramanian M, Goel N, et al. A2A adenosine receptor may allow expansion of T cells lacking effector functions in extracellular adenosine-rich microenvironments. Journal of Immunology. 2009;**183**(9):5487-5493

[98] Gessi S, Varani K, Merighi S, Cattabriga E, Avitabile A, Gavioli R, et al. Expression of A3 adenosine receptors in human lymphocytes: Up-regulation in T cell activation. Molecular Pharmacology. 2004;**65**(3):711-719

[99] Sluyter R, Barden JA, Wiley JS. Detection of P2X purinergic receptors on human B lymphocytes. Cell and Tissue Research. 2001;**304**(2):231-236

[100] Przybyła T, Sakowicz-Burkiewicz M, Pawełczyk T. Purinergic signaling in B cells. Acta Biochimica Polonica. 2018;**65**(1):1-7

[101] Sengstake S, Boneberg E-M, Illges H. CD21 and CD62L shedding are both inducible via P2X7Rs. International Immunology. 2006;**18**(7):1171-1178

[102] Pupovac A, Geraghty NJ, Watson D, Sluyter R. Activation of the P2X7 receptor induces the rapid shedding of CD23 from human and murine B cells. Immunology and Cell Biology. 2015;**93**(1):77-85

[103] Sakowicz-Burkiewicz M, Kocbuch K, Grden M, Maciejewska I, Szutowicz A, Pawelczyk T. High glucose concentration impairs ATP outflow and immunoglobulin production by human peripheral B lymphocytes: Involvement of P2X7 receptor. Immunobiology. 2013;**218**(4):591-601

[104] Conter LJ, Song E, Shlomchik MJ, Tomayko MM. CD73 expression is dynamically regulated in the germinal center and bone marrow plasma cells are diminished in its absence. PLoS One. 2014;**9**(3):e92009

[105] Cao F, Hu L-Q, Yao S-R, Hu Y, Wang D-G, Fan Y-G, et al. P2X7 receptor:

A potential therapeutic target for autoimmune diseases. Autoimmunity Reviews. 2019;**18**(8):767-777

[106] Cojocaru M, Cojocaru IM, Silosi I, Vrabie CD. Manifestations of systemic lupus erythematosus. Maedica. 2011;**6**(4):330-336

[107] Yaniv G, Twig G, Shor DB, Furer A, Sherer Y, Mozes O, et al. A volcanic explosion of autoantibodies in systemic lupus erythematosus: A diversity of 180 different antibodies found in SLE patients. Autoimmunity Reviews. 2015;**14**(1):75-79

[108] Rai G, Rai R, Saeidian AH, Rai M. Microarray to deep sequencing: Transcriptome and miRNA profiling to elucidate molecular pathways in systemic lupus erythematosus. Immunologic Research. 2016;**64**(1):14-24

[109] Rai R, Chauhan SK, Singh VV, Rai M, Rai G. Heat shock protein 27 and its regulatory molecules express differentially in SLE patients with distinct autoantibody profiles. Immunology Letters. 2015;**164**(1):25-32

[110] Darrah E, Andrade F. NETs: The missing link between cell death and systemic autoimmune diseases? Frontiers in Immunology. 2012;**3**:428

[111] Magna M, Pisetsky DS. The role of cell death in the pathogenesis of SLE: Is pyroptosis the missing link? Scandinavian Journal of Immunology. 2015;**82**(3):218-224

[112] Chauhan SK, Rai R, Singh VV, Rai M, Rai G. Differential clearance mechanisms, neutrophil extracellular trap degradation and phagocytosis, are operative in systemic lupus erythematosus patients with distinct autoantibody specificities. Immunology Letters. 2015;**168**(2):254-259

[113] Mahajan A, Herrmann M, Muñoz LE. Clearance deficiency and cell death pathways: A model for the pathogenesis of SLE. Frontiers in Immunology. 2016;**7**:35

[114] Di Virgilio F, Giuliani AL. Purinergic signalling in autoimmunity: A role for the P2X7R in systemic lupus erythematosus? Biomedical Journal. 2016;**39**(5):326-338

[115] Kahlenberg JM, Carmona-Rivera C, Smith CK, Kaplan MJ. Neutrophil extracellular trap-associated protein activation of the NLRP3 inflammasome is enhanced in lupus macrophages. Journal of Immunology. 2013;**190**(3):1217-1226

[116] Wang D, Drenker M, Eiz-Vesper B, Werfel T, Wittmann M. Evidence for a pathogenetic role of interleukin-18 in cutaneous lupus erythematosus. Arthritis and Rheumatism. 2008;**58**(10):3205-3215

[117] Hu D, Liu X, Chen S, Bao C. Expressions of IL-18 and its binding protein in peripheral blood leukocytes and kidney tissues of lupus nephritis patients. Clinical Rheumatology. 2010;**29**(7):717-721

[118] Kahlenberg JM, Thacker SG, Berthier CC, Cohen CD, Kretzler M, Kaplan MJ. Inflammasome activation of IL-18 results in endothelial progenitor cell dysfunction in systemic lupus erythematosus. Journal of Immunology. 2011;**187**(11):6143-6156

[119] Turner CM, Tam FW, Lai PC, Tarzi RM, Burnstock G, Pusey CD, et al. Increased expression of the pro-apoptotic ATP-sensitive P2X7 receptor in experimental and human glomerulonephritis. Nephrology, Dialysis, Transplantation. 2007;**22**(2):386-395

[120] Zhao J, Wang H, Dai C, Wang H, Zhang H, Huang Y, et al. P2X7 blockade

attenuates murine lupus nephritis by inhibiting activation of the NLRP3/ASC/caspase 1 pathway. Arthritis and Rheumatism. 2013;**65**(12):3176-3185

[121] Taylor SR, Turner CM, Elliott JI, McDaid J, Hewitt R, Smith J, et al. P2X7 deficiency attenuates renal injury in experimental glomerulonephritis. The Journal of the American Society of Nephrology. 2009;**20**(6):1275-1281

[122] Hu S, Yu F, Ye C, Huang X, Lei X, Dai Y, et al. The presence of P2RX7 single nuclear polymorphism is associated with a gain of function in P2X7 receptor and inflammasome activation in SLE complicated with pericarditis. Clinical and Experimental Rheumatology. 2020;**38**(3):442-449

[123] Portales-Cervantes L, Niño-Moreno P, Salgado-Bustamante M, García-HernándezMH,Baranda-CandidoL, Reynaga-Hernández E, et al. The His155Tyr (489C>T) single nucleotide polymorphism of P2RX7 gene confers an enhanced function of P2X7 receptor in immune cells from patients with rheumatoid arthritis. Cellular Immunology. 2012;**276**(1-2):168-175

[124] Li M, Yang C, Wang Y, Song W, Jia L, Peng X, et al. The expression of P2X7 receptor on Th1, Th17, and regulatory T cells in patients with systemic lupus erythematosus or rheumatoid arthritis and its correlations with active disease. Journal of Immunology. 2020;**205**(7):1752-1762

[125] Portales-Cervantes L, Niño-Moreno P, Doníz-Padilla L, Baranda-Candido L, García-Hernández M, Salgado-Bustamante M, et al. Expression and function of the P2X7 purinergic receptor in patients with systemic lupus erythematosus and rheumatoid arthritis. Human Immunology. 2010;**71**(8):818-825

[126] Faliti CE, Gualtierotti R, Rottoli E, Gerosa M, Perruzza L, Romagnani A, et al. P2X7 receptor restrains pathogenic Tfh cell generation in systemic lupus erythematosus. The Journal of Experimental Medicine. 2019;**216**(2):317-336

[127] Le Gall SM, Legrand J, Benbijja M, Safya H, Benihoud K, Kanellopoulos JM, et al. Loss of P2X7 receptor plasma membrane expression and function in pathogenic B220+ double-negative T lymphocytes of autoimmune MRL/lpr mice. PLoS One. 2012;**7**(12):e52161

[128] He Y, Gallman AE, Xie C, Shen Q, Ma J, Wolfreys FD, et al. P2RY8 variants in lupus patients uncover a role for the receptor in immunological tolerance. The Journal of Experimental Medicine. 2022;**219**(1):e20211004

[129] Knight JS, Mazza LF, Yalavarthi S, Sule G, Ali RA, Hodgin JB, et al. Ectonucleotidase-mediated suppression of lupus autoimmunity and vascular dysfunction. Frontiers in Immunology. 2018;**9**:1322

[130] Hesse J, Siekierka-Harreis M, Steckel B, Alter C, Schallehn M, Honke N, et al. Profound inhibition of CD73-dependent formation of anti-inflammatory adenosine in B cells of SLE patients. eBioMedicine. 2021;**73**:103616

[131] Aletaha D, Smolen JS. Diagnosis and management of rheumatoid arthritis: A review. Journal of the American Medical Association. 2018;**320**(13):1360-1372

[132] Idzko M, Ferrari D, Eltzschig HK. Nucleotide signalling during inflammation. Nature. 2014;**509**(7500):310-317

[133] Bhagavatham SKS, Kannan V, Darshan VMD, Sivaramakrishnan V. Nucleotides modulate synoviocyte proliferation and osteoclast differentiation in macrophages with

potential implications for rheumatoid arthritis. 3. Biotech. 2021;**11**(12):504

[134] Zhang WJ, Luo HL, Zhu ZM. The role of P2X4 receptors in chronic pain: A potential pharmacological target. Biomedicine & Pharmacotherapy. 2020;**129**:110447

[135] Tsuda M, Kuboyama K, Inoue T, Nagata K, Tozaki-Saitoh H, Inoue K. Behavioral phenotypes of mice lacking purinergic P2X4 receptors in acute and chronic pain assays. Molecular Pain. 2009;**5**:28

[136] Hamoudi C, Zhao C, Abderrazak A, Salem M, Fortin PR, Sévigny J, et al. The purinergic receptor P2X4 promotes Th17 activation and the development of arthritis. Journal of Immunology. 2022;**208**(5):1115-1127

[137] Li F, Guo N, Ma Y, Ning B, Wang Y, Kou L. Inhibition of P2X4 suppresses joint inflammation and damage in collagen-induced arthritis. Inflammation. 2014;**37**(1):146-153

[138] Lopez-Castejon G, Theaker J, Pelegrin P, Clifton AD, Braddock M, Surprenant A. P2X(7) receptor-mediated release of cathepsins from macrophages is a cytokine-independent mechanism potentially involved in joint diseases. Journal of Immunology. 2010;**185**(4):2611-2619

[139] Fan Z-D, Zhang Y-Y, Guo Y-H, Huang N, Ma H-H, Huang H, et al. Involvement of P2X7 receptor signaling on regulating the differentiation of Th17 cells and type II collagen-induced arthritis in mice. Scientific Reports. 2016;**6**:35804

[140] Gao F, Li X. P2Y11 receptor antagonist NF340 ameliorates inflammation in human fibroblast-like synoviocytes: An implication in rheumatoid arthritis. IUBMB Life. 2019;**71**(10):1552-1560

[141] Orriss IR, Wang N, Burnstock G, Arnett TR, Gartland A, Robaye B, et al. The P2Y6 receptor stimulates bone resorption by osteoclasts. Endocrinology. 2011;**152**(10):3706-3716

[142] Su X, Floyd DH, Hughes A, Xiang J, Schneider JG, Uluckan O, et al. The ADP receptor P2RY12 regulates osteoclast function and pathologic bone remodeling. The Journal of Clinical Investigation. 2012;**122**(10):3579-3592

[143] Lazarowski ER, Harden TK. UDP-sugars as extracellular signaling molecules: Cellular and physiologic consequences of P2Y14 receptor activation. Molecular Pharmacology. 2015;**88**(1):151-160

[144] Stamp LK, Hazlett J, Roberts RL, Frampton C, Highton J, Hessian PA. Adenosine receptor expression in rheumatoid synovium: A basis for methotrexate action. Arthritis Research & Therapy. 2012;**14**(3):R138

[145] Knowles HJ. The adenosine a(2B) receptor drives osteoclast-mediated bone resorption in hypoxic microenvironments. Cell. 2019;**8**(6):624

[146] Moncrieffe H, Ursu S, Pesenacker A, Gordon-Smith S, Zheng D, Wedderburn L. Autoimmune susceptibility gene critically influences CD39 T cell expression and function in modulating human inflammation (P3313). The Journal of Immunology. 2013;**190**(1 Supplement):175.174

[147] da Silva JLG, Passos DF, Bernardes VM, Leal DBR. ATP and adenosine: Role in the immunopathogenesis of rheumatoid arthritis. Immunology Letters. 2019;**214**:55-64

[148] Botta Gordon-Smith S, Ursu S, Eaton S, Moncrieffe H, Wedderburn LR.

Correlation of low CD73 expression on synovial lymphocytes with reduced adenosine generation and higher disease severity in juvenile idiopathic arthritis. Arthritis & Rhematology. 2015;**67**(2):545-554

[149] Sadovnick AD, Gu BJ, Traboulsee AL, Bernales CQ, Encarnacion M, Yee IM, et al. Purinergic receptors P2RX4 and P2RX7 in familial multiple sclerosis. Human Mutation. 2017;**38**(6):736-744

[150] van Olst L, Rodriguez-Mogeda C, Picon C, Kiljan S, James RE, Kamermans A, et al. Meningeal inflammation in multiple sclerosis induces phenotypic changes in cortical microglia that differentially associate with neurodegeneration. Acta Neuropathologica. 2021;**141**(6):881-899

[151] Dziedzic A, Miller E, Saluk-Bijak J, Niwald M, Bijak M. The molecular aspects of disturbed platelet activation through ADP/P2Y(12) pathway in multiple sclerosis. International Journal of Molecular Sciences. 2021;**22**(12):657

[152] Vincenzi F, Corciulo C, Targa M, Merighi S, Gessi S, Casetta I, et al. Multiple sclerosis lymphocytes upregulate A2A adenosine receptors that are antiinflammatory when stimulated. European Journal of Immunology. 2013;**43**(8):2206-2216

[153] Álvarez-Sánchez N, Cruz-Chamorro I, Díaz-Sánchez M, Lardone PJ, Guerrero JM, Carrillo-Vico A. Peripheral CD39-expressing T regulatory cells are increased and associated with relapsing-remitting multiple sclerosis in relapsing patients. Scientific Reports. 2019;**9**(1):2302

[154] Jakovljevic M, Lavrnja I, Bozic I, Milosevic A, Bjelobaba I, Savic D, et al. Induction of NTPDase1/CD39 by reactive microglia and macrophages is associated with the functional state during EAE. Frontiers in Neuroscience. 2019;**13**:410

[155] Domercq M, Zabala A, Matute C. Purinergic receptors in multiple sclerosis pathogenesis. Brain Research Bulletin. 2019;**151**:38-45

[156] Sidoryk-Węgrzynowicz M, Strużyńska L. Astroglial and microglial purinergic P2X7 receptor as a major contributor to neuroinflammation during the course of multiple sclerosis. International Journal of Molecular Sciences. 2021;**22**(16):8404

[157] Villani AC, Sarkizova S, Hacohen N. Systems immunology: Learning the rules of the immune system. Annual Review of Immunology. 2018;**36**:813-842

[158] Eberhardt N, Bergero G, Mazzocco Mariotta YL, Aoki MP. Purinergic modulation of the immune response to infections. Purinergic Signal. 2022:**18**(1):93-113

[159] Costales MG, Alam MS, Cavanaugh C, Williams KM. Extracellular adenosine produced by ecto-5′-nucleotidase (CD73) regulates macrophage pro-inflammatory responses, nitric oxide production, and favors Salmonella persistence. Nitric Oxide. 2018;**72**:7-15

[160] Raczkowski F, Rissiek A, Ricklefs I, Heiss K, Schumacher V, Wundenberg K, et al. CD39 is upregulated during activation of mouse and human T cells and attenuates the immune response to Listeria monocytogenes. PLoS One. 2018;**13**(5):e0197151

[161] Chiacchio T, Casetti R, Butera O, Vanini V, Carrara S, Girardi E, et al. Characterization of regulatory T cells identified as CD4(+)CD25(high) CD39(+) in patients with active

tuberculosis. Clinical and Experimental Immunology. 2009;**156**(3):463-470

[162] Théâtre E, Frederix K, Guilmain W, Delierneux C, Lecut C, Bettendorff L, et al. Overexpression of CD39 in mouse airways promotes bacteria-induced inflammation. Journal of Immunology. 2012;**189**(4):1966-1974

[163] Savio LEB, de Andrade MP, Figliuolo VR, de Avelar Almeida TF, Santana PT, Oliveira SDS, et al. CD39 limits P2X7 receptor inflammatory signaling and attenuates sepsis-induced liver injury. Journal of Hepatology. 2017;**67**(4):716-726

[164] Amaral EP, Ribeiro SC, Lanes VR, Almeida FM, de Andrade MR, Bomfim CC, et al. Pulmonary infection with hypervirulent mycobacteria reveals a crucial role for the P2X7 receptor in aggressive forms of tuberculosis. PLoS Pathogens. 2014;**10**(7):e1004188

[165] Li X, Kondo Y, Bao Y, Staudenmaier L, Lee A, Zhang J, et al. Systemic adenosine triphosphate impairs neutrophil chemotaxis and host defense in sepsis. Critical Care Medicine. 2017;**45**(1):e97-e104

[166] Taylor JM, Han Z. Purinergic receptor functionality is necessary for infection of human hepatocytes by hepatitis delta virus and hepatitis B virus. PLoS One. 2010;**5**(12):e15784

[167] Pacheco PA, Faria RX, Ferreira LG, Paixão IC. Putative roles of purinergic signaling in human immunodeficiency virus-1 infection. Biology Direct. 2014;**9**:21

[168] Corrêa G, de ALC, Fernandes-Santos C, Gandini M, Petitinga Paiva F, Coutinho-Silva R, et al. The purinergic receptor P2X7 role in control of dengue virus-2 infection and cytokine/chemokine production in infected human monocytes. Immunobiology. 2016;**221**(7):794-802

[169] Hasan D, Shono A, van Kalken CK, van der Spek PJ, Krenning EP, Kotani T. A novel definition and treatment of hyperinflammation in COVID-19 based on purinergic signalling. Purinergic Signal. 2022;**18**(1):13-59

[170] Nikolova M, Carriere M, Jenabian MA, Limou S, Younas M, Kök A, et al. CD39/adenosine pathway is involved in AIDS progression. PLoS Pathogens. 2011;**7**(7):e1002110

[171] Schuler PJ, Macatangay BJ, Saze Z, Jackson EK, Riddler SA, Buchanan WG, et al. $CD4^+CD73^+$ T cells are associated with lower T-cell activation and C reactive protein levels and are depleted in HIV-1 infection regardless of viral suppression. AIDS. 2013;**27**(10):1545-1555

[172] Tóth I, Le AQ, Hartjen P, Thomssen A, Matzat V, Lehmann C, et al. Decreased frequency of CD73+CD8+ T cells of HIV-infected patients correlates with immune activation and T cell exhaustion. Journal of Leukocyte Biology. 2013;**94**(4):551-561

[173] Séror C, Melki MT, Subra F, Raza SQ, Bras M, Saïdi H, et al. Extracellular ATP acts on P2Y2 purinergic receptors to facilitate HIV-1 infection. The Journal of Experimental Medicine. 2011;**208**(9):1823-1834

[174] Orellana JA, Velasquez S, Williams DW, Sáez JC, Berman JW, Eugenin EA. Pannexin1 hemichannels are critical for HIV infection of human primary CD4+ T lymphocytes. Journal of Leukocyte Biology. 2013;**94**(3):399-407

[175] Freeman TL, Swartz TH. Purinergic receptors: Elucidating the role of these

immune mediators in HIV-1 fusion. Viruses. 2020;**12**(3):290

[176] Giroud C, Marin M, Hammonds J, Spearman P, Melikyan GB. P2X1 receptor antagonists inhibit HIV-1 fusion by blocking virus-coreceptor interactions. Journal of Virology. 2015;**89**(18):9368-9382

[177] Manzoor S, Akhtar U, Naseem S, Khalid M, Mazhar M, Parvaiz F, et al. Ionotropic purinergic receptors P2X4 and P2X7: Proviral or antiviral? An insight into P2X receptor signaling and hepatitis C virus infection. Viral Immunology. 2016;**29**(7):401-408

[178] Tsai CY, Liong KH, Gunalan MG, Li N, Lim DS, Fisher DA, et al. Type I IFNs and IL-18 regulate the antiviral response of primary human γδ T cells against dendritic cells infected with dengue virus. Journal of Immunology. 2015;**194**(8):3890-3900

[179] Zarei M, Sahebi Vaighan N, Ziai SA. Purinergic receptor ligands: The cytokine storm attenuators, potential therapeutic agents for the treatment of COVID-19. Immunopharmacology and Immunotoxicology. 2021;**43**(6):633-643

[180] Leão Batista Simões J, Fornari Basso H, Cristine Kosvoski G, Gavioli J, Marafon F, Elias Assmann C, et al. Targeting purinergic receptors to suppress the cytokine storm induced by SARS-CoV-2 infection in pulmonary tissue. International Immunopharmacology. 2021;**100**:108150

[181] Simões JLB, Bagatini MD. Purinergic signaling of ATP in COVID-19 associated Guillain-Barré syndrome. Journal of Neuroimmune Pharmacology. 2021;**16**(1):48-58

[182] Caillon A, Trimaille A, Favre J, Jesel L, Morel O, Kauffenstein G. Role of neutrophils, platelets, and extracellular vesicles and their interactions in COVID-19-associated thrombopathy. Journal of Thrombosis and Haemostasis. 2022;**20**(1):17-31

[183] Schultz IC, Bertoni APS, Wink MR. Purinergic signaling elements are correlated with coagulation players in peripheral blood and leukocyte samples from COVID-19 patients. Journal of Molecular Medicine (Berlin, Germany). 2022:**100**(4):569-584

[184] Ahmadi P, Hartjen P, Kohsar M, Kummer S, Schmiedel S, Bockmann JH, et al. Defining the CD39/CD73 axis in SARS-CoV-2 infection: The CD73(−) phenotype identifies polyfunctional cytotoxic lymphocytes. Cell. 2020;**9**(8):1750

[185] Ribeiro DE, Oliveira-Giacomelli Á, Glaser T, Arnaud-Sampaio VF, Andrejew R, Dieckmann L, et al. Hyperactivation of P2X7 receptors as a culprit of COVID-19 neuropathology. Molecular Psychiatry. 2021;**26**(4):1044-1059

[186] Di Virgilio F, Tang Y, Sarti AC, Rossato M. A rationale for targeting the P2X7 receptor in coronavirus disease 19. British Journal of Pharmacology. 2020;**177**(21):4990-4994

Chapter 3

Purinergic Signaling in Covid-19 Disease

Hailian Shen

Abstract

SARS-CoV-2 virus infection causes the Covid-19 disease pandemic. Purinergic signaling is a form of extracellular signaling. Purinergic signaling plays significant role in the pathology of Covid-19. Purinergic system includes extracellular nucleotides, nucleosides, ectonucleotidases, and purinergic receptors. ATP, ADP, and adenosine are the main nucleotides, nucleosides. CD39 and CD73 are the main ectonucleotidases. There are two classes of purinergic receptors, P1 and P2. Each of them can be further divided, P1 into A1, A2A, A2B, and A3, P2 into P2X, and P2Y. In Covid-19, the purinergic system is disordered. SARS-CoV-2 viruses invading leads to extracellular ATP and ADP accumulation, purinergic receptor abnormally activation, tissue homeostasis balance is broken, which lead to inflammation even hyperinflammation with cytokine storm and thrombosis et al. symptoms. Currently, Covid-19 therapeutic medicine is still in shortage. Target purinergic system components is a promising way to treat Covid-19, which will help inhibit inflammation and prevent thrombosis. Currently, many relevant preclinical and clinical trials are ongoing. Some are very promising.

Keywords: purinergic system, purinergic signaling, purinergic receptor, SARS-CoV-2, Covid-19

1. Introduction

Coronavirus disease 2019 (Covid-19) is caused by severe acute respiratory syndrome coronavirus 2 (SARS-CoV-2) [1]. The first Covid-19 case was reported at the end of December 2019 in Wuhan, China [2]. From then on, the virus transmission was swiftly spread leading to the Covid-19 pandemic. So far, more than 6 billion people have infected SARS-CoV-2 virus worldwide, and more than 6million people died in the pandemic. Although now there are several kinds of vaccines have been put into use and been proven to be safe, effective and life-saving, the pandemic does not stop. There are several reasons may be responsible for this. The first one, not 100 percent people inoculate with vaccines. Second, like all the other vaccines, the Covid-19 vaccines do not fully protect everyone who is vaccinated, namely the efficacy rates of the vaccines are less than 100%. The SARS-CoV-2 is enveloped positive-chain RNA virus, which is prone to mutate. New mutation strains more easily escape from the vaccine defense. Third, vaccine protection is time-dependent, not lifelong. The SARS-CoV-2 virus transmission is through hand-mouth-eye contact and infected droplets released by coughing and sneezing. After the virus enters the body, it will combine with the cell surface angiotensin-converting enzyme 2 (ACE2) receptor through its envelope spike protein [3]. With the help of transmembrane serine protease2 (TMPRSS2), the virus will enter successfully into the cell [4]. There are some other cell-surface proteins, like Eph receptors, Neuropilin 1, and CD147 et al. which can also act as SARS-CoV-2 virus cell entry helpers [5, 6]. From asymptomatic to life-threatening acute respiratory distress syndrome (ARDS), the manifestations of the SARS-CoV-2 virus-infected people are quite variant. The most common symptoms found in the clinical presentation are fatigue, anosmia, ageusia, dizziness, headache, obtundation, myalgia, diarrhea, anorexia, fever, cough, pneumonia, and dyspnea [7, 8]. In severe cases, the virus infection will cause hyper inflammation with cytokine storm and also thrombogenesis can occur [9].

Purinergic signaling plays a pivotal role in SARS-CoV-2 virus infection, participates in the regulation of the innate immune system and platelet function, which are highly relevant for hemostasis, inflammatory, and thrombosis processes [10]. The purinergic system may be a possible target for SARS-CoV-2 treatments [11]. In this chapter, the purinergic signaling in Covid-19 disease will be introduced.

2. Purinergic system

Purinergic signaling is a form of extracellular signaling. In 1972, Geoffrey Burnstock proposed that adenosine triphosphate (ATP) can act as a neurotransmitter, which opened up a new area [12]. Purinergic system is composed of extracellular nucleotides, nucleosides, ectonucleartides, and purinergic receptors.

2.1 Nucleotides and nucleosides

The nucleotides and nucleosides are mainly ATP, adenosine diphosphate (ADP), and adenosine (Ado). In addition, Uridine diphosphate (UDP), uracil-diphosphate-sugar (UDP -sugar), and Nicotinamide adenine dinucleotide (NAD+) can also act as purinergic signal molecules [13]. Besides as a purinergic signal molecule, ATP is the universal currency of energy metabolism. 10% ATP is produced in cytoplasm by glycolysis. In this way, 1 molecular glucose lysis can generate 2 molecular ATPs. While the 90% ATP is synthesized in mitochondria through Krebs cycle. 1 molecular glucose oxidative-phosphorylation can create 30 ATPs. The Krebs cycle needs oxygen. As ACE2 receptors are highly expressed in lung, SARS-CoV-2 virus very easily invades this tissue leading to lower ventilation rates in Covid-19 patients. Hypoxia will cause ATP production inefficiency to lead to low intracellular ATP [14, 15]. The situation will be further intensified by ATP going outside the cell through Pannexin-1 channel (PANX1). PANX1 expression is increasing in Covid 19. Normally, extracellular ATP concentration is very low, commonly the concentration is less than 3 nanomolar, while intracellular one can be high at millimolar (2–8 mM). So, Covid-19 patients' extracellular ATP level is dramatically increased, along with extracellular ADP accumulation. SARS-CoV-2 virus-infected cell lysis is another important reason for high extracellular ATP. Extracellular ATP catalyzed by CD39 (ectonucleoside triphosphate diphosphohydrolase-1, ENTPD1) to dephosphorylated into ADP, further into AMP which is still by CD39. Then CD73 (ecto-5′-nucleotidase, NT5E) will fully convert AMP to Ado. From ATP to Ado, this pathway is called canonical adenosinergic pathway, which was firstly illustrated by Yegutkin et al. Ado can be produced through a non-canonical alternative pathway, which starts from NAD+. NAD+ first is metabolized into ADP-ribose (ADPR) by CD38, then into AMP by CD203a, and further into Ado by CD73. So, two pathways converge on CD73. The life of Ado is very short. It will soon be deaminized into inosine by adenosine deaminase (ADA). Inosine is much more stable. Ado can also be taken into the cell through equilibrative nucleoside transporter-1 (ENT1) [16].

2.2 Ectonucleartidases

Ectonucleotidases are nucleotide metabolizing enzymes, which located on cytoplasmic membrane. The main function of Ectonucleotidases is to catalyze nucleotides hydrolyzation to balance nucleotides and neucleosides. Ectonucleotidases can be classed into 4 families: the ectonucleoside triphosphate diphosphohydrolases (NTPDase1-4,8), the ectonucleotide pyrophosphatase phosphodiesterases (NPP1-3), ecto-5′-nucleotidase and alkaline phosphatase (**Table 1**). CD39 is type 1 ecto-nucleoside triphosphate diphosphohydrolase, while CD73 is ecto-5′-nucleotidase. CD39 and CD73 not only regulate AMP existing, but also GMP state [17]. CD39 was overexpressed in COVID-19 patients' plasma and some immune cell subsets and related to hypoxemia [18]. Plasma soluble form of CD39 (sCD39) was related to length of hospital stay and independently associated with intensive care unit admission. Soluble Plasma CD39 may be used to predict covid-19 patients' clinical prognosis, which is suggested as a promising biomarker for COVID-19 severity. CD39 is a defined marker of exhausted T cell [19]. T cell exhaustion and dysfunction are hallmarks of severe COVID-19. Both $CD4^+$ Tim-3^+ $CD39^+$ T cell and $CD8^+$ Tim-3^+ $CD39^+$ T cell significant increase in multiple tissues, like lung, liver, spleen and PBMCs, of critical covid 19 patients [20]. CD39 expression was also found up-regulated in plasmablasts. CD39 higher expression was also reported in the placenta of a 23 year old woman of pregnancy complicated by SARS-CoV-2 virus infection and the accompanying placental complement C4d deposition [21]. Regulatory T (Treg) cells have been shown to play an essential role in immune homeostasis in many diseases and pathological conditions [22]. Some studies have reported that $CD4^+CD25^+CD39^+$ Tregs have more immunosuppressive effects than $CD4^+CD25^+CD39^-$ Tregs [23]. In covid-19, $CD39^+$ Tregs are decreased in juvenile patients in an age-dependent manner while in adult patients,$CD39^+$ Tregs increased with disease severity [24]. However, CD73 expression is down-regulated in plasmablasts, $CD8^+$ T cells and natural killer T cells (NKT)[25]. But there is one paper shows that moderate and severe cases have increased expression of CD39 and CD73 in total leukocytes. CD38, a catalytic case of non-canonical adenosinergic pathway mentioned above, is upregulated.

2.3 Purinergic receptors

Purinergic receptors can be separated into P1 and P2 [26]. P1 receptor can be further subclassed into A1, A2A, A2B and A3, P2 receptor further into P2X and P2Y. P1 and P2Y are G protein-coupled metabolic receptors, while P2X receptors are fast ligand gated ion channel.

P1 receptor is also known as adenosine receptor as its endogenous ligand is Ado. Caffeine and theophylline are two best known P1 receptor antagonists [27, 28]. Once Ado binding to P1 receptor, the conformation of the receptor will change to activate the coupled G protein. Activated G protein will cause intracellular cyclic adenosine

Family	Members	Hydrolysis pathways	—	—
NTPDases	NTPDase 1	ATP → ADP + P_i	ADP → AMP + P_i	—
—	NTPDase 2	ATP → ADP + P_i	ADP → AMP + P_i	—
—	NTPDase 3	ATP → ADP + P_i	ADP → AMP + P_i	—
—	NTPDase 4	UDP → UMP + P_i	UTP → UDP + P_i	—
—	NTPDase 5*	UDP → UMP + P_i	GDP → GMP + P_i	—
—	NTPDase 6*	GDP → GMP + P_i		—
—	NTPDase 7*	UTP → UDP + P_i	GTP → GDP + P_i	—
—	NTPDase 8	ATP → ADP + P_i	ADP → AMP + P_i	—
NPPs	NPP1	ATP → AMP + 2P_i	ADP → AMP + P_i	3′,5′-cAMP → AMP
—	NPP2	ATP → AMP + 2Pi	ATP → ADP + P_i	ADP → AMP + P_i
—		GTP → GDP + P_i	3′,5′-cAMP → AMP	AMP → adenosine + P_i
—	NPP3	ATP → AMP + 2P_i	ADP → AMP + P_i	3′,5′-cAMP → AMP
Ecto-5′-nucleotidase		AMP → adenosine + P_i		—
Alkaline phosphatase	—	ATP → ADP + P_i	ADP → AMP + P_i	AMP → adenosine + P_i

**Intracellular enzymes.*
Adapt from Seldin and Giebisch, The Kidney Physiological &Pathophysiological, 4th Edition, Oct1, 2007.

Table 1.
Major hydrolysis pathways of Ectonucleotidases.

monophosphate (cAMP) level change through acting on adenylate-cyclase. 4 subtype P1 receptors in human, encoded by different genes, interact with different G protein subunits. A1 and A3 receptors preferentially bind to inhibitory regulative Gi/o proteins, inhibiting adenylate-cyclase and cyclic AMP production, whereas the receptors of A2 family are generally coupled to stimulative regulative Gs protein that trigger intracellular cAMP accumulation. A1 and A2A are high affinity receptors, while A2B and A3 are low affinity receptors. Mitogen-activated protein kinase signaling pathway has also been reported to be another P1 receptor downstream pathway. In immune system, A1 receptor is mainly expressed on Neutrophils and immature dendritic cells, A2A on most immune cells, A2B on macrophages and dendritic cells, while A3 on neutrophils and mast cells. Ado binding to A1 receptor will produce chemotaxis function. On the contrary, A3 receptor activation will reduce neutrophil and stimulate mast cells degranulation. A2A and A2B receptors evoke immune suppress. The well-known anti-inflammatory effects of Ado are mediated by these two receptors [29]. A2B expression is upregulated in Covid 19.

P2 receptor can be sub-grouped into P2X and P2Y. P2X receptors belong to a larger family of receptors known as the ENaC/P2X superfamily. They are homologs. Structurally, P2X receptors and ENaC are very similar. PX2 receptors have a wide tissue distribution, which being expressed in nervous system, the pulmonary and digestive systems, muscle, bone, and immune system et al. Blood cells, like red cells, lymphocytes and macrophages and platelets, can be traced to have P2X receptors' expression. P2X receptor family contains 7 members, P2X1 to P2X7 respectively, which are heterotrimers or homotrimers. Another name of P2X7 receptor is P2Z. ATP is the full agonist of

P2X receptors, which can activate all P2X family receptors. NAD+ is also an activator of P2X receptors. However, there are some nucleotide-specific variations between these two ligands. For example, among the P2X receptor subtypes, the P2X7 receptor is unique in facilitating the induction of nonselective pores that allow entry of organic cations and dye molecules. Upon stimulation with ATP. As little as 100 μM ATP was sufficient to activate the nonselective pore, whereas NAD+ at concentrations up to 2 mM had no effect. The affinities between ATP and different P2X subgroups are also different. ADP and AMP, when purified, are inactive at P2X receptors. Activation of P2X receptors leads to influx of cations such as sodium and calcium, and further to depolarize the excitable cells. Among the 7 P2X receptors, P2X7 which mainly expressed on Macrophages, mast cells, microglia, pancreas, skin, and endocrine organs, is most studied and plays a pivotal role in SARS-CoV-2 virus infection associated inflammation, which is a promising target to treat Covid-19 disease [30]. Beside to P2X7, P2X1, P2X4, and P2X5 have been detected increasing expression in Covid 19 patients either.

P2Y receptors are seven-transmembrane proteins belonging to the class A family of G protein-coupled receptors (GPCRs), which are the δ group of rhodopsin-like GPCRs [31]. Structurally, P2Y receptors are characterized by extracellular N-terminal, which followed by seven hydrophobic transmembrane (7-TM) α-helices (TM-1 to TM-7) connected by three extracellular loops (ECL) and three intracellular loops, and ending in an intracellular C-terminus. An ECL serves to bind the receptor ligand(s), while intracellular regions mediate G protein activation and participate in P2Y receptor regulation. P2Y receptors can form both homodimers and heterodimers to further increase the biochemical and pharmacological spectrum of P2YRs. P2Y receptor's family consists of 8 subunits, P2Y [4, 5, 14, 18, 28, 30, 32, 33]. The gaps of the subunit numbers are because of the fact that the assignment of numbers to certain putative P2Y receptors was later shown to be premature, with some of the previously designated sequences being P2Y species homologs and others being other types of receptors. P2Y receptors are present in almost all human tissues, where they exert various biological functions based on their G-protein coupling. Different from P2X receptors which have only ATP and NAD^+ two native nucleotide agonists, P2Y receptors respond not only to nucleotides (ATP, ADP, UTP, UDP, NAD^+) but also to nucleotide sugars such as UDP-glucose. According to the G-protein coupling difference, P2Y can be classed into Gq-coupled, P2Y1-like receptors (P2Y1, P2Y2, P2Y4, P2Y6 and P2Y11) and Gi -coupled, P2Y12-like receptors (P2Y12, P2Y13 and P2Y14) (**Table 2**). As P1 receptors, when ligands combine to P2Y receptors, the conformations of the receptors will change to transfer the signal to the coupled G-proteins. Heteromeric G-proteins (Gαβγ) will dissociate into Gα subunits and Gβγ complexes, which activate or regulate downstream effector pathways. P2Y11 is the only P2Y member which can activate cAMP pathway. P2Y receptors also play very important roles in in SARS-CoV-2 virus infection related inflammation and can be served as therapeutic targets. P2RY1 and P2RY12 have been shown to be elevated in Covid-19.

2.4 Roles of purinergic signaling in the inflammation of Covid-19

The inflammation of Covid-19 is the most important biological response of the body tissue to SARS-CoV-2 invasion [34]. Usually inflammation is the innate immune protective response involving immune cells, blood vessels, and molecular mediators. However, in severe covid-19 patients, it can develop into hyperinflammation, which can be life-threatening. Hyperinflammation is thought to be the base to develop into severe Covid 19. About hyperinflammation, currently there is no clear-cut definition.

groups	Receptors	Tissue distribution	Agonist(s)	G Protein coupling
P2Y1-like	P2Y1	Brain, epithelial and endothelial cells, platelets, immune cells, osteoclasts	ADP, NAD^+	Gq
—	P2Y2	Immune cells, epithelial and endothelial cells, kidney tubules, osteoblasts	ATP ≈ UTP	Gq-Gi
—	P2Y4	endothelial cells, placenta	UTP	Gq-Gi
—	P2Y6	airway and intestinal epithelial cells, spleen, placenta, T-cells, thymus,	UDP	Gq
—	P2Y11	spleen, intestine, granulocytes, macrophage	ATP, NAD^+	Gq-Gs
P2Y12-like	P2Y12	Platelets, brain (glial cells), microglial cells	ADP	Gi
—	P2Y13	Spleen, brain, lymph nodes, bone marrow	UDP-Sugar	Gi
—	P2Y14	Placenta, mast cells, adipose tissue, stomach, intestine, discrete brain regions	UDP	Gi

Table 2.
Human P2Y receptors.

The criteria of hyperinflammation are not consistent. Most people think the condition of hyperinflammation as a form of very severe inflammation with cytokine storm which is out of tissue homeostatic control to lead to ARDS or other organs failure. Purinergic signal system is in the pivotal position of pro-inflammation and anti-inflammation axis. Once the balance is broken, pro-inflammation factors being far more than the ones of anti-inflammation, hyperinflammation will happen.

SARS-CoV-2 invasion leads to extracellular ATP and ADP accumulation. ATP will bind to P2X7 receptor. Though P2X7 receptor expressed on almost all type human and mouse cells, the levels of the ones on monocyte and macrophage are much higher. ATP binding to P2X7 receptor leads to pore forming on the cell surface to cause K^+ efflux. Intracellular K^+ depletion and extracellular K^+ concentration increase is necessary and sufficient to activate and assembly the NLR family pyrin domain containing 3 (NLRP3) inflammasome to promote proteolytic cleavage, maturation and secretion of pro-inflammatory cytokines interleukin 1β (IL-1β) and interleukin 18 (IL-18) [30]. So far there is no evidence to show that SARS-CoV-2 can directly activate NLRP3 inflammasome. Flowing P2X7 receptor activation other cytokines and chemokines, for example, IL-6, TNF-α, CCL2, IL-8, CCL3 and CXCL2, of pro-fibrotic factors such as TGF-β, and extracellular matrix remodeling factors, for example, metalloproteinase-9 and tissue inhibitor of metalloproteinase (TIMP)-1 will also be released. In mild Covid 19, the extracellular ATP concentration lower than 100uM, after proinflammation process starts the anti-inflammation response will be triggered either. First reaction is CD39 / CD73 will convert ATP into Ado. Ado will activate P1 receptor. As above mentioned, A2A and A2B receptors activation will launch immune suppress. At the same time, activation of A2A receptor will promote the differentiation of naïve T-cells towards regulatory T-cells (Tregs). Treg will secrete immune suppressive factors, like IL-10 and TGF-β, to restrict immune reaction. However, when extracellular ATP concentration over 100uM, the situation will become worse, dramatic immune response will lead to sever inflammation. If extracellular ATP concentration is over 1 mM, hyperinflammation will be inevitable in most cases. High amount extracellular ATP accumulation leads to prolong P2X7 receptor activation. P2x7 receptor overactivation leads to macropore formation and cytolysis with uncontrolled ATP outgoing and cytokines release. What making the situation even worse is the anti-inflammation process being out of control. P2X7 receptor activation inhibits he suppressive potential and stability of Tregs. Tregs clonal proliferation and mature are suppressed, Treg death increasing. Treg depletion leads to IL10 et al. immune suppressive factors drop. In severe covid-19 patient, the expression of forkhead box protein P3 (FoxP3), a marker of Treg, was monitored lower than that in health control. On the other side, CD73 express is down-regulated, which blocks the production of Ado, which cause P1 receptors desensitization [35]. So, though A2B is detected to have higher expression in Covid-19, it is less activated. The homeostatic out-control at last results in the hyperinflammation exploding. Not only P2X7 receptor play a role in Covid-19 inflammation, other P2 receptors also have functions in the proinflammation. For ex, ATP-P2X4, ADP-P2Y6, ADP-P2Y12, and UDP-sugure-P2Y14 et al. mediated signaling all can stimulate inflammation via actions on innate immune cells, especially dendritic cells and macrophages.

2.5 Roles of purinergic signaling in the thrombosis of Covid-19

In addition to inflammation, many Covid-19 patients also have microvascular thrombosis, which have been confirmed by autopsy. Clinical detection has also provided very solid evidence. Covid-19 patients have high level of circulating D-dimeris (a fibrin/fibrinogen degradation product), prolonged prothrombin time, upregulated expression of tissue factor (TF, encoded by F3 gene) et al [32]. The Covid-19 related stroke incidence was reported increase, either. Purinergic system not only participate in inflammation but also involved in thrombosis, it is like a bridge to connect the two processes. The complex interplay between the two processes is described as thromoinflammation [36, 37].

Thrombosis is the formation of a blood clot inside a blood vessel, obstructing the flow of blood through the circulatory system. In Covid-19 disease the balance of coagulation and fibrinolysis is broken, which leads to thrombosis happen [38]. Blood coagulation can be divided into two pathways: intrinsic pathway and extrinsic pathway. Intrinsic cascade starts from blood contacts the damaged blood vessel surface or other high molecular surface with negative charges to induce factor XII activation, which following by factor XI and activation. On the phospholipid surface of the activated platelet, factor IX together with factor VIII (vW factor) and Ca^{2+} will activate factor X. Extrinsic pathway, which also called tissue factor pathway, is beginning from factor VII being activated by tissue factor. Activated factor VII can directly activate factor X. So, two coagulation pathways converge on factor X activation. Activated factor collaborated with factor V and Ca^{2+}catalyzes prothrombin to become thrombin. Also, the phospholipid surface of the activated platelet is necessary for this reaction. Thrombin will continue to catalyze fibrinogen to convert into fibrin. Purinergic system can promote coagulation from several aspects [33]. Activated platelet plays very important roles in blood coagulation. ADP can directly activate platelet. As above mentioned, Covid-19 patients have extracellular ADP accumulation. The accumulated ADP can bind to P2Y12 receptor located on the surface of the platelet to activate it. The activated platelet will secrete more ADP and vW factor et al. ADP can also activate platelet through combine to P2Y1 receptor. ATP is also a platelet activator.

ATP can interact with platelet P2X1 receptor. ATP binds to macrophage P2X7 receptor enhance tissue factor expression and release to trigger extrinsic pathway [39]. The third way is that purinergic signal can stimulate neutrophils activation to produce reactive oxygen (ROS). The overwhelming production of ROS can result the release of neutrophil extracellular traps (NETs), which is web-like structures composed of chromatin containing neutrophil granule proteins [40]. NETs can further activate factor XII to activate the intrinsic pathway [19].

3. Therapeutic targets

Regarding the important roles purinergic signal plays in Covid-19 disease, the members of purinergic system have been used as therapeutic targets to reduce morbidity and mortality.

3.1 Adenosine and Ado metabolism enzymes

As above described, Ado can exert anti-inflammation effects through active P1 receptor [41, 42]. Clinically, Ado is used in cardiac diseases diagnosis and treatment. Preclinically, Ado administration was demonstrated to be able to attenuate lung injury [43]. Ado being reported can also be applied in Covid-19 patient treatment. A patient suffering from SARS-CoV-2-related ARDS on routine therapies who did not show clinical improvements, inhaled adenosine in a mixture of 21% oxygen was applied. After 5 days, the SARS-CoV-2 test was negative and a rapid improvement in clinical condition as well as radiological pictures were shown. The main concern about Ado used in disease treatment is its short half-life in vivo. In the future more stable Ado analogs may be developed. Ado metabolism enzymes blocking methods is another way to elevate extracellular Ado. Pentostatin (2′ deoxycoformycin) and EHNA (erytho-9-(2-hydroxy-3-nonyl) adenine hydrochloride) are two ADA inhibitors. Clinically, pentostatin is used in Hairy Cell Leukemia treatment. It is suggested pentostatin might be beneficial in late-stage ARDS. Not like pentostatin which only inhibits ADA enzyme activity, EHNA can also bind to P1 receptors and adenosine deaminase complexing protein 2(CD26). EHNA potentially has anticancer effects, but so far has been used clinically. EHNA is also suggested to be potentially used in Covid-19 therapy. Dipyridamole (DIP) is a ENT1 inhibitor, which can prevent extracellular Ado uptake [44]. DIP is an approved antiplatelet drug, clinically being used to prevent stroke, and being proved to have high safety [45]. The bleeding risk of DIP is similar to that of aspirin. Currently, three clinical trials evaluating efficacy of dipyridamole for the treatment of COVID-19 have been registered (identifiers: NCT04424901; NCT04391179; NCT04410328) [46, 47]. Apart from anticoagulant and anti-inflammatory effects, it is speculated that DIP can also blunt SARS-CoV-2 replication.

3.2 CD39

As CD39 plays very important roles in extracellular ATP and ADP hydrolysis, its expression and activity closely related to inflammation and thrombosis [48]. Several approaches have been attempted to target CD39. One of them is using soluble CD39 to antithrombosis [48]. However, this method easily causes bleeding. To overcome this side effect, new strategies is worked out. The core of these new strategies is to link the recombinant soluble CD39 to other molecules, like PSGL-1, the receptor for P selectin on leukocyte surface, and single chain antibody (scFV) specific against GPIIb/IIIa, the platelet fibrinogen receptor, and glycoprotein VI (GPVI) Fc fusion protein et al.

3.3 P1 receptors

P1 receptor family contain1 A1, A2A, A2B, A3. These 4 receptors have different functions [49]. Activated A2A and A2B can suppress immune response. As mentioned above that caffeine and theophylline are two best known P1 receptors antagonists. Both are non-selective antagonists except for A3 receptor, they can inhibit the other 3 P1 receptors, namely A1, A2A, and A2B, at therapeutic concentrations. Theophylline is more potent. As these receptors have different functions in inflammation, inhibit these three receptors will have different effects. For example, theophylline has been shown to have both proinflammatory and anti-inflammatory effects [50]. The latter one might be stronger. Recently, shown by preclinical data that theophylline can potentially amplify the anti-inflammatory effect of corticosteroids and reduce corticosteroid resistance. Now one clinical trial, which theophylline is designed to be nasally administrated to treat the Covid-19 patients who have been received intranasal and oral corticosteroids, is on-going (Identifier: NCT04789499). However, there is one report sharing that theophylline treatment induced sinus bradycardia in two cases of Covid-19 patients [51]. Pentoxifylline (PTX) can active A2A receptor. When PTX binds to A2A receptor, it will stimulate to secrete IL-10 et al. immune suppressive molecules to inhibit inflammation. PTX treatment is also shown to help reduce IL-6 serum concentration, as well as diminish IL-1b level. PTX has been recommended to be applied in Covid-19 therapy.

3.4 X2P7 receptor

X2P7 receptor is the most important pro- inflammation purinergic receptor. X2P7 receptor block is predicated to be able to ameliorate inflammation. X2P7 receptor activation can also potentially induce of VEGF release. P2X7 receptor blockade can inhibit VEGF-dependent increase in vascular permeability, and therefor prevent lung oedema. Several X2P7 receptor antagonists are suggested to be used in Covid-19 therapy [27, 52]. Colchicine is one of such inhibitors. Colchicine is a tricyclic lipid-soluble alkaloid extracted from Colchicum autumnalle and *gloriosa superba*. Colchicine is a well-known of microtubule polymerization inhibitor, which in the early time was found to be able to block cell mitosis. Hereafter, its anti-inflammation effects was revealed. Colchicine has been clinically used as an anti-inflammatory agent for long-term treatment of Behçet's disease and also used to treat many other diseases, like pericarditis, pulmonary fibrosis, biliary cirrhosis, various vasculitides, pseudogout, spondyloarthropathies, calcinosis, scleroderma, and amyloidosis et al. Colchicine not only can inhibit X2P7 receptor, but also can block X2P2 receptor pathway. Several clinical trials have shown that colchicine can limit the production of some cytokines, like IL-1b, IL-18, and IL-6 et al., of Covid-19 patients. NIH has included colchicine in Covid-19 treatment guideline.

Lidocaine is another P2X7 receptor antagonist, which routinely used as local anesthesia in clinic. It's readily available and affordable. Recently, a clinical trial (Identifier:NCT04609865) is carrying out in a French group, in which lidocaine is intravenously administrated to treat Covid-19 disease [6]. However, the half-maximal effective concentration (IC50) for P2X7R inhibition of lidocaine is much higher than the maximal tolerable plasma concentration where adverse effects start to develop. A Peru group modified the protocol. 28 (three mild, 21 moderate and four severe) COVID-19 patients were treated with 0.5% lidocaine HCL solution with an intravenous dose of 1 mg/kg once a day for 2 days and 2% lidocaine HCL solution with a subcutaneous dose of 1 mg/kg once a day for 2 days, which results in the improvement in pain, cough, respiratory rate and oxygen saturation. Another group directly carried out subdermal administration of lidocaine in 6 critical ill Covid-19 induced ARDS patients. The author claimed that although all six patients appeared to respond positively to the treatment and no severe adverse effects were observed, no final conclusions could be made on the efficacy of lidocaine in critically ill COVID-19 patients.

3.5 P2Y12 receptor

P2Y12 receptor is the main purinergic receptor responsible for SARS-CoV-2 virus related thrombosis, Theoretically, P2Y12 receptor blocking can confine thrombosis of Covid-19, and also can curb the inflammation. Several P2Y12 antagonists (clopidorgel, prasugrel, ticagrelor and cangrelor) have been clinically used to prevent thrombosis in patients at risk of heart attack for about 20 years [27, 53]. Recently, these antagonists have been reevaluated for its effects in Covid-19 related thrombosis treatment. Several clinical trials (Identifiers are NCT04505774; NCT04409834, and NCT04333407.) are on the way. One of them, titled "accelerating Covid-19 therapeutic and vaccines 4 acute (ACTIV-4A, Identifier: NCT04505774)" has been finished and reported [54]. This is a randomized clinical trial, which aims to test if P2Y12receptor antagonists can enhance heparin therapeutic effects in mild Covid-19 patients. The answer is no. The results demonstrated that among non- critical ill hospitalized Covid-19 patients, the use of a P2Y12 receptor antagonist in addition to a therapeutic dose of heparin, compared with a therapeutic dose of heparin only, did not result in an increased odds of improvement in organ support– free days within 21 days during hospitalization. However, as the author mentioned that this trial tested only the combination of a P2Y12 inhibitor with anticoagulant therapy, it remains possible that use of a P2Y12 inhibitor as a sole antithrombotic agent may improve outcomes in patients with COVID-19. In addition, the potential for benefit with a longer treatment duration or at an earlier stage of illness (before hospitalization) cannot be ruled out.

4. Conclusion

Purinergic signal is involved in SARS-CoV-2 viruses causing Covid-19, which plays a pivotal role in the pathology of Covid-19 disease. Purinergic signal participates in the regulation of the innate immune system and platelet function et al., which are highly relevant for hemostasis, inflammatory and thrombosis processes. SARS-CoV-2 virus infection will lead to the abnormality of purinergic system to break the body homeostasis further to inflammation and thrombosis. Purinergic system components have been suggested to be Covid-19 therapeutic targets. Currently many preclinical and clinical trials have been in progress to test this hypothesis. Promising data have brought new hope to the patients.

Author details

Hailian Shen
University of Texas Health Science Center at San Antonio, San Antonio, USA

*Address all correspondence to: shenh4@uthscsa.edu

References

[1] Rat P, Olivier E, Dutot M. SARS-CoV-2 vs. SARS-CoV-1 management: Antibiotics and inflammasome modulators potential. European Review for Medical and Pharmacological Sciences. 2020;**24**(14):7880-7885

[2] Dos Anjos F, Simões JLB, Assmann CE, Carvalho FB, Bagatini MD. Potential therapeutic role of purinergic receptors in cardiovascular disease mediated by SARS-CoV-2. Journal of Immunology Research. 2020:8632048

[3] Doğan HO, Şenol O, Bolat S, Yıldız ŞN, Büyüktuna SA, Sarıismailoğlu R, et al. Understanding the pathophysiological changes via untargeted metabolomics in COVID-19 patients. Journal of Medical Virology. 2021;**93**(4):2340-2349

[4] Dietl P, Frick M. Channels and transporters of the pulmonary lamellar body in health and disease. Cell. 2021;**11**(1):45

[5] Zalpoor H, Akbari A, Samei A, Forghaniesfidvajani R, Kamali M, Afzalnia A, et al. The roles of Eph receptors, neuropilin-1, P2X7, and CD147 in COVID-19-associated neurodegenerative diseases: Inflammasome and JaK inhibitors as potential promising therapies. Cellular & Molecular Biology Letters. 2022;**27**(1):10

[6] Ribeiro DE, Oliveira-Giacomelli Á, Glaser T, Arnaud-Sampaio VF, Andrejew R, Dieckmann L, et al. Hyperactivation of P2X7 receptors as a culprit of COVID-19 neuropathology. Molecular Psychiatry. 2021;**26**(4): 1044-1059

[7] Edwards C, Klekot O, Halugan L, Korchev Y. Follow your nose: A key clue to understanding and treating COVID-19. Frontiers in Endocrinolology. 2021;**12**:747744

[8] Simões JLB, Bagatini MD. Purinergic signaling of ATP in COVID-19 associated Guillain-Barre Syndrome. Journal of Neuroimmune Pharmacology. 2021;**16**(1):48-58

[9] Hasan D, Shono A, van Kalken CK, van der Spek PJ, Krenning EP, Kotani T. A novel definition and treatment of hyperinflammation in COVID-19 based on purinergic signalling. Purinergic Signal. 2022;**18**(1):13-59

[10] Franciosi MLM, Lima MDM, Schetinger MRC, Cardoso AM. Possible role of purinergic signaling in COVID-19. Molecular and Cellular Biochemistry. 2021;**476**(8):2891-2898

[11] Pacheco PAF, Faria RX. The potential involvement of P2X7 receptor in COVID-19 pathogenesis: A new therapeutic target? Scandinavian Journal of Immunology. 2021;**93**(2):e12960

[12] Franco R, Rivas-Santisteban R, Lillo J, Camps J, Navarro G, Reyes-Resina I. 5-Hydroxytryptamine, glutamate, and ATP: Much more than neurotransmitters. Frontiers in Cell and Development Biology. 2021;**9**:667815

[13] Grahnert A, Klein C, Hauschildt S. Involvement of P2X receptors in the NAD(+)-induced rise in [Ca (2+)] (i) in human monocytes. Purinergic Signal. 2009;**5**(3):309-319

[14] Liu W, Zhu X, Mozneb M, Nagahara L, Hu TY, Li CZ. Lighting up ATP in cells and tissues using a simple aptamer-based fluorescent probe. Mikrochimica Acta. 2021;**188**(10):352

[15] Abraham EH, Guidotti G, Rapaport E, Bower D, Brown J, Griffin RJ, et al. Cystic fibrosis improves COVID-19 survival and provides clues for treatment of SARS-CoV-2. Purinergic Signal. 2021;**17**(3):399-410

[16] Schultz IC, Bertoni APS, Wink MR. Purinergic signaling elements are correlated with coagulation players in peripheral blood and leukocyte samples from COVID-19 patients. Journal of Molecular Medicine. 2022;**29**:1

[17] Wu D, Shu T, Yang X, Song JX, Zhang M, Yao C, et al. Plasma metabolomic and lipidomic alterations associated with COVID-19. National Science Review. 2020;7(7):1157-1168

[18] Díaz-García E, García-Tovar S, Alfaro E, Zamarrón E, Mangas A, Galera R, et al. Role of CD39 in COVID-19 Severity: Dysregulation of purinergic signaling and thromboinflammation. Frontiers in Immunology. 2022;**13**:847894

[19] Wang N, Vuerich M, Kalbasi A, Graham JJ, Csizmadia E, Manickas-Hill ZJ, et al. Limited TCR repertoire and ENTPD1 dysregulation mark late-stage COVID-19. iScience. 2021;**24**(10):103205

[20] Shahbazi M, Moulana Z, Sepidarkish M, Bagherzadeh M, Rezanejad M, Mirzakhani M, et al. Pronounce expression of Tim-3 and CD39 but not PD1 defines CD8 T cells in critical Covid-19 patients. Microbial Pathogenesis. 2021;**153**:104779

[21] Shimao Y, Yamauchi A, Ohtsuka T, Terao K, Kodama Y, Yamada N, et al. C4d deposition and CD39 downregulation in the placental infection by SARS-CoV-2. Pathology International. 2022

[22] Setoguchi R, Hori S, Takahashi T, Sakaguchi S. Homeostatic maintenance of natural Foxp3(+) CD25(+) CD4(+) regulatory T cells by interleukin (IL)-2 and induction of autoimmune disease by IL-2 neutralization. The Journal of Experimental Medicine. 2005;**201**(5):723-735. DOI: 10.1084/jem.20041982

[23] Bastid J, Cottalorda-Regairaz A, Alberici G, Bonnefoy N, Eliaou JF, Bensussan A. ENTPD1/CD39 is a promising therapeutic target in oncology. Oncogene. 2013;**32**(14):1743-1751. DOI: 10.1038/onc.2012.269

[24] Simsek A, Kizmaz MA, Cagan E, Dombaz F, Tezcan G, Asan A, et al. Assessment of CD39 expression in regulatory T-cell subsets by disease severity in adult and juvenile COVID-19 cases. Journal of Medical Virology. 2022

[25] Wildner NH, Ahmadi P, Schulte S, Brauneck F, Kohsar M, Lütgehetmann M, et al. B cell analysis in SARS-CoV-2 versus malaria: Increased frequencies of plasmablasts and atypical memory B cells in COVID-19. Journal of Leukocyte Biology. 2021;**109**(1):77-90

[26] Alves VS, Leite-Aguiar R, Silva JPD, Coutinho-Silva R, Savio LEB. Purinergic signaling in infectious diseases of the central nervous system. Brain, Behavior, and Immunity. 2020;**89**:480-490

[27] Zarei M, Sahebi Vaighan N, Ziai SA. Purinergic receptor ligands: The cytokine storm attenuators, potential therapeutic agents for the treatment of COVID-19. Immunopharmacology and Immunotoxicology. 2021 Dec;**43**(6):633-643

[28] Simões JLB, de Araújo JB, Bagatini MD. Anti-inflammatory therapy by cholinergic and purinergic modulation in multiple sclerosis associated with SARS-CoV-2 infection. Molecular Neurobiology. 2021;**58**(10):5090-5111

[29] Gonzales JN, Gorshkov B, Varn MN, Zemskova MA, Zemskov EA, Sridhar S, et al. Protective effect of adenosine receptors against lipopolysaccharide-induced acute lung injury. American Journal of Physiology. Lung Cellular and Molecular Physiology. 2014;**306**(6):L497-L507

[30] Leão Batista Simões J, Fornari Basso H, Cristine Kosvoski G, Gavioli J, Marafon F, Elias Assmann C, et al. Targeting purinergic receptors to suppress the cytokine storm induced by SARS-CoV-2 infection in pulmonary tissue. International Immunopharmacology. 2021;**100**:108150

[31] Klaver D, Thurnher M. Control of Macrophage Inflammation by P2Y Purinergic Receptors. Cell. 2021;**10**(5):1098

[32] Brun JF, Varlet-Marie E, Myzia J, Raynaud de Mauverger E, Pretorius E. Metabolic influences modulating erythrocyte deformability and eryptosis. Metabolites. 2021;**12**(1):4

[33] Ziegler O, Sriram N, Gelev V, Radeva D, Todorov K, Feng J, et al. The cardiac molecular setting of metabolic syndrome in pigs reveals disease susceptibility and suggests mechanisms that exacerbate COVID-19 outcomes in patients. Scientific Reports. 2021;**11**(1):19752

[34] Wauters E, Van Mol P, Garg AD, Jansen S, Van Herck Y, Vanderbeke L, et al. Discriminating mild from critical COVID-19 by innate and adaptive immune single-cell profiling of bronchoalveolar lavages. Cell Research. 2021;**31**(3):272-290

[35] Ahmadi P, Hartjen P, Kohsar M, Kummer S, Schmiedel S, Bockmann JH, et al. Defining the CD39/CD73 axis in SARS-CoV-2 infection: The CD73(-) phenotype identifies polyfunctional cytotoxic lymphocytes. Cells. 2020;**9**(8):1750

[36] Lintzmaier Petiz L, Glaser T, Scharfstein J, Ratajczak MZ, Ulrich H. P2Y14 receptor as a target for neutrophilia attenuation in severe COVID-19 cases: From hematopoietic stem cell recruitment and chemotaxis to thrombo-inflammation. Stem Cell Reviews and Reports. 2021;**17**(1):241-252

[37] Sriram K, Insel PA. Inflammation and thrombosis in COVID-19 pathophysiology: Proteinase-activated and purinergic receptors as drivers and candidate therapeutic targets. Physiological Reviews. 2021;**101**(2):545-567

[38] Edwards C. New horizons: Does mineralocorticoid receptor activation by cortisol cause ATP release and COVID-19 complications? The Journal of Clinical Endocrinology and Metabolism. 2021;**106**(3):622-635

[39] Furlan-Freguia C, Marchese P, Gruber A, Ruggeri ZM, Ruf W. P2X7 receptor signaling contributes to tissue factor-dependent thrombosis in mice. The Journal of Clinical Investigation. 2011;**121**(7):2932-2944

[40] Caillon A, Trimaille A, Favre J, Jesel L, Morel O, Kauffenstein G. Role of neutrophils, platelets, and extracellular vesicles and their interactions in COVID-19-associated thrombopathy. Journal of Thrombosis and Haemostasis. 2022 Jan;**20**(1):17-31

[41] Garcia-Dorado D, García-del-Blanco B, Otaegui I, Rodríguez-Palomares J, Pineda V, Gimeno F, et al. Intracoronary injection of adenosine before reperfusion in patients with ST-segment elevation myocardial infarction: A randomized controlled

clinical trial. International Journal of Cardiology. 2014;**177**(3):935-941

[42] Jin Z, Duan W, Chen M, Yu S, Zhang H, Feng G, et al. The myocardial protective effects of adenosine pretreatment in children undergoing cardiac surgery: A randomized controlled clinical trial. European Journal of Cardiothoracic Surgery. 2011;**39**(5):e90-e96

[43] Lu Q, Harrington EO, Newton J, Casserly B, Radin G, Warburton R, et al. Adenosine protected against pulmonary edema through transporter- and receptor A2-mediated endothelial barrier enhancement. American Journal of Physiology. Lung Cellular and Molecular Physiology. 2010;**298**(6):L755-L767

[44] Kanthi Y, Knight JS, Zuo Y, Pinsky DJ. New (re)purpose for an old drug: Purinergic modulation may extinguish the COVID-19 thromboinflammatory firestorm. JCI Insight. 2020;**5**(14):e140971

[45] Liu X, Li Z, Liu S, Sun J, Chen Z, Jiang M, et al. Potential therapeutic effects of dipyridamole in the severely ill patients with COVID-19. Acta Pharmaceutica Sinica B. 2020;**10**(7):1205-1215

[46] Li X, Berg NK, Mills T, Zhang K, Eltzschig HK, Yuan X. Adenosine at the Interphase of Hypoxia and Inflammation in Lung Injury. Frontiers in Immunology. 2021;**11**:604944

[47] Cardoso AM. COVID-19 and purinergic signaling: The need for investigation. Purinergic Signal. 2020;**16**(3):451-452

[48] Morello S, Caiazzo E, Turiello R, Cicala C. Thrombo-inflammation: A focus on NTPDase1/CD39. Cell. 2021;**10**(9):2223

[49] Abouelkhair MA. Targeting adenosinergic pathway and adenosine A(2A) receptor signaling for the treatment of COVID-19: A hypothesis. Medical Hypotheses. 2020;**144**:110012

[50] Sawalha K, Habash FJ, Vallurupalli S, Paydak H. Theophylline in Treatment of COVID-19 Induced Sinus Bradycardia. Clinical Practice. 2021;**11**(2):332-336

[51] Di Nicolantonio JJ, Barroso-Aranda J. Harnessing adenosine A2A receptors as a strategy for suppressing the lung inflammation and thrombotic complications of COVID-19: Potential of pentoxifylline and dipyridamole. Medical Hypotheses. 2020;**143**:110051

[52] Di Virgilio F, Tang Y, Sarti AC, Rossato M. A rationale for targeting the P2X7 receptor in Coronavirus disease 19. British Journal of Pharmacology. 2020;**177**(21):4990-4994

[53] Pereira NL, Avram R, So DY, Iturriaga E, Byrne J, Lennon RJ, et al. Rationale and design of the TAILOR-PCI digital study: Transitioning a randomized controlled trial to a digital registry. American Heart Journal. 2021;**232**: 84-93

[54] Berger JS, Kornblith LZ, Gong MN, Reynolds HR, Cushman M, Cheng Y, et al. ACTIV-4a Investigators. Effect of P2Y12 inhibitors on survival free of organ support among non-critically ill hospitalized patients with COVID-19: A Randomized Clinical Trial. JAMA. 2022;**327**(3):227

Section 2

Purinergic System and Disease

Chapter 4

Graft-Versus-Host Disease: Pathogenesis and Treatment

Shin Mukai

Abstract

Graft-versus-host disease (GVHD) is a disabling complication after allogeneic hematopoietic stem cell transplantation (HSCT) and negatively impacts patients' quality of life. GVHD is classified into 2 forms according to clinical manifestations. Acute GVHD (aGVHD) typically affects the skin, gastrointestinal tract, and liver, whereas chronic GVHD occurs systemically and shows diverse manifestations similar to autoimmune diseases such as eosinophilic fasciitis, scleroderma-like skin disease. GVHD is induced by complicated pathological crosstalk between immune cells of the host and donor and involves various signaling pathways such as purinergic signaling. Although the past several decades have seen significant progress in the understanding of mechanisms of GVHD and several drugs have been approved by FDA for the prevention and treatment of GVHD, there is still vast scope for improvement in the therapy for GVHD. Thus, new drugs for GVHD will need to be developed. Towards this goal, this chapter succinctly summarises the pathogenic process of GVHD and emerging GVHD treatments in order to provide some insights into the mechanisms of GVHD and facilitate the development of novel drugs.

Keywords: inflammation, fibrosis, therapeutic targets, drug development

1. Introduction

Graft-versus-host disease (GVHD) is a debilitating complication that can determine the prognosis of allogeneic hematopoietic stem cell transplantation (HSCT) and subject 40–60% of HSCT recipients to a risk of death and disability [1]. GVHD is composed of acute GVHD (aGVHD) and chronic GVHD (cGVHD). For the classification of the 2 types of GVHD, the classifier should be clinical manifestations instead of time after HSCT [2]. However, in many cases, aGVHD appears within 100 days after HSCT and causes severe inflammation mostly in the skin, gastrointestinal tract, and liver [3]. cGVHD generally occurs systemically 6 months or later after HSCT, and its symptoms are similar to those of autoimmune diseases [4]. Complex interactions between donor and host immune cells are implicated in the pathogenesis of GVHD. It is thought that aGVHD is induced primarily by donor T cells' cytotoxic responses against host tissues through recognition of host polymorphic histocompatibility antigens [5]. On the other hand, the mechanisms of cGVHD are more complicated and still poorly understood [6]. Although the use of corticosteroids alone or in combination with immunosuppressive agents is the recommended first-line strategy for the treatment of GVHD, its efficacy is not satisfactory [3, 7]. The prevalence of allogeneic HSCT for the treatment of hematologic diseases has increased the need for the development of efficacious second-line therapies which can mitigate symptoms of GVHD without compromising a graft-versus-leukemia effect, where donor T cells eliminate host leukemia cells. To date, various signaling pathways and pathogenic events in the context of GVHD have been intensively investigated. As a result, several FDA-approved drugs for GVHD have recently emerged. This chapter concisely summarises therapeutic targets and newly emerging drugs for the 2 forms of GVHD with the goal to facilitate the development of novel GVHD treatments for human use.

2. Clinical manifestations of GVHD

aGVHD can occur after the engraftment of donor-derived cells in the transplant recipient [8]. Symptoms of aGVHD can develop within weeks after the transplantation [9]. It has been believed that aGVHD can primarily affect the skin, gastrointestinal (GI) tract, and/or liver [10]. HSCT recipients can manifest rash, increased bilirubin, diarrhea, and vomiting [11]. Most recently, mounting evidence suggests that other organs such as the central nervous system, lungs, ovaries and testis, thymus, bone marrow, and kidney can be susceptible to aGVHD [12].

Clinical manifestations of cGVHD are different from those of aGVHD. The onset of cGVHD can be divided into the following 3 cases: (1) occurring when aGVHD is present, (2) emerging after a period of resolution from aGVHD, and (3) developing de novo [13]. Immune dysregulation and absence of functional tolerance are characteristic of cGVHD, and symptoms of cGVHD are reminiscent of those of autoimmune disorders [13]. Clinical presentations of cGVHD can be as follows: (i) rash, raised or discolored areas, skin thickening or tightening, (ii) dry eye or vision changes, (iii) dry mouth, white patches inside the mouth, (iv) diarrhea and weight loss, (v) shortness of breath due to lung disorders and (vi) abnormal liver function [14]. It was challenging for clinicians to reach an agreement on the diagnosis, the timing of treatment, and how to grade cGVHD [15]. In order to overcome these difficulties, the National Institute of Health (NIH) consensus created diagnostic criteria for cGVHD in 2005 and revised the criteria in 2014 [16, 17]. The authors considered the severity of involvement of the skin, mouth, eyes, gastrointestinal tract, liver, lungs, joint fascia, and genital tract in order to define manifestations of cGVHD in its target organs and establish a scoring system.

Corticosteroids are used with or without immunosuppressive drugs as the first-line therapy for aGVHD and cGVHD in clinical settings [3, 7, 18, 19]. However, approximately 50% of patients who receive steroid therapy will be resistant to it, although mechanisms of steroid resistance remain to be elucidated [3, 7, 18, 19]. In addition, corticosteroid therapies also cause various undesired effects such as diabetes, obesity, osteoporosis, hypertension, glaucoma, and liver damage [20]. Thus, medical settings are in need of effective treatments of steroid-refractory aGVHD and cGVHD [3, 7, 18, 19].

3. General GVHD biology

GVHD has a complex pathophysiology, which initially begins with damage to host tissues by chemotherapy and radiation therapy before allogeneic HSCT (**Figure 1**) [21].

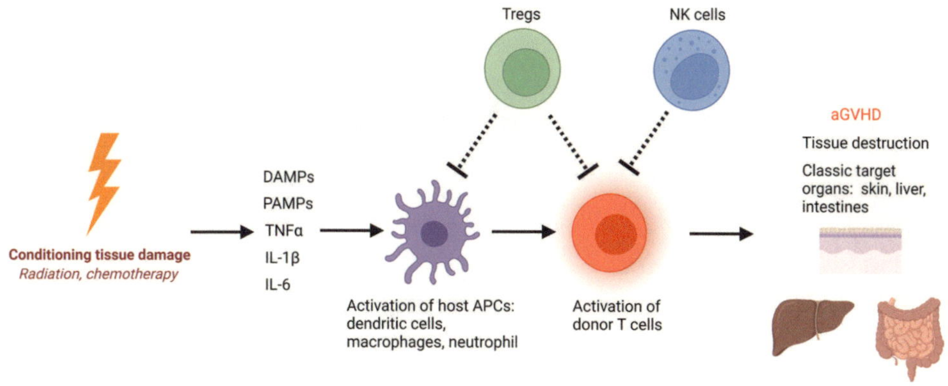

Figure 1.
The overview of aGVHD pathogenesis. The preconditioning regimen causes tissue damage. It generates DAMPs, PAMPs and proinflammatory cytokines such as TNFα, IL-1β and IL-6, which activates host APCs. The activated APCs present antigens to donor T cells, and the activated T cells infiltrate aGVHD target organs and produce an excessive amount of IFNγ and IL-17, leading to abnormal inflammation and tissue damage. This figure is created with BioRender.

Due to this, damage-associated molecular patterns (DAMPs), pathogen-associated molecular patterns (PAMPs), and inflammatory cytokines are released [22]. These stimuli activate host dendritic cells (DCs), leading to the expression of major histocompatibility complex class I (MHC-I) and class II (MHC-II) on the host DCs [22]. The mature host DCs activate donor-derived T cells in the graft [22]. The activated donor T cells migrate to aGVHD-susceptible organs and promote the excessive production of pro-inflammatory cytokines such as interferon (IFN)-γ and interleukin (IL)-17 [23, 24]. It results in abnormal inflammation and tissue damage [23, 24]. While it is believed that donor-derived $CD4^+$ and $CD8^+$ T cells play a pivotal role in mediating aGVHD [25], several other types of immune cells are reportedly involved in the pathogenic process of aGVHD [26]. Neutrophils contribute to the development of intestinal aGVHD [27]. A previous report suggests that neutrophils in the ileum migrate to mesenteric lymph nodes, presenting antigens on their MHC-II and promoting T cell expansion [28]. Donor monocyte-derived macrophages with potent immunological functions are implicated in the pathophysiology of cutaneous aGVHD by secreting chemokines, stimulating T cells, and mediating direct cytotoxicity [29, 30]. In contrast, regulatory T cells (Tregs) are thought to serve a suppressive role in aGVHD without significantly reducing the graft-versus-leukemia (GVL) effect [31, 32]. Recent reports suggest that donor-derived natural killer (NK) cells can have an inhibitory effect in aGVHD by promoting the depletion of allo-reactive T cells while showing the GVL effect [33]. A recent study indicates that the occurrence and severity of aGVHD could be associated with the disordered reconstitution of $CD56^{high}$ NK cells [34].

While mechanisms of cGVHD are still incompletely understood, recent evidence suggests that there are several observations characteristic of cGVHD (**Figure 2**) [35]. The thymus is damaged due to the conditioning regimen and/or the prior occurrence of aGVHD, leading to impaired negative selection of alloreactive $CD4^+$ T cells [36]. Alloreactive T cells are activated by antigen-presenting cells (APCs), resulting in their expansion and polarization toward type 1, type 2, and type 17 helper T (Th1, Th2, and Th17) cells [35]. These immune deviations lead to the production of proinflammatory and profibrotic inflammatory cytokines such as IFNγ, IL-6, IL-17, IL-4, and transforming growth factor β (TGFβ), which skew macrophages and fibroblasts towards

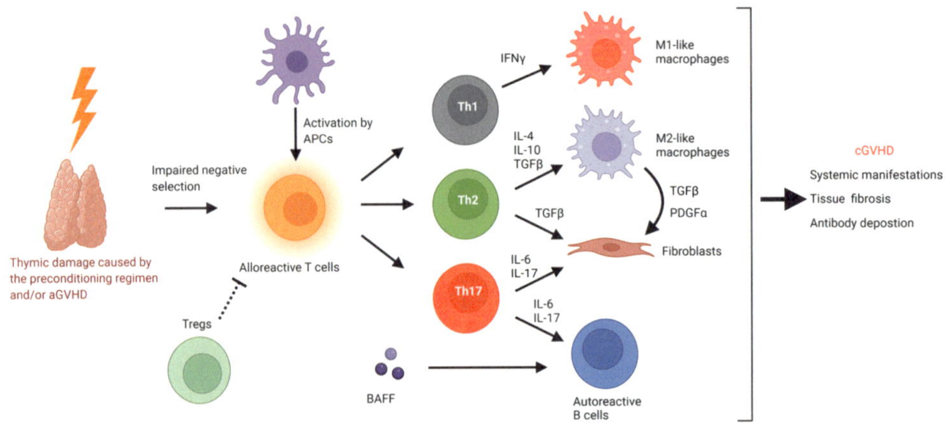

Figure 2.
Overview of cGVHD pathogenesis. The thymus is damaged due to the preconditioning regimen and/or aGVHD. Due to the damage, the negative selection of alloreactive T cell is impaired. Alloreactive T cells are polarised into Th1, Th2 or Th17 cells. Th1 cells produce IFNγ, which drives macrophages to an M1-like phenotype to promote inflammation. IL-4, IL-10 and TGFβ produced by Th2 cells facilitate macrophage differentiation into an M2-like phenotype. Activation and proliferation of tissue fibroblasts are induced by (i) TGFβ from Th2 cells, (ii) PDGFα and TGFβ from M2-like macrophages and (iii) IL-6 and IL-17 from Th17 cells, leading to collagen production and fibrosis. B cells are activated by IL-6 and IL-17 from Th17 cells, and the alloreactivity of B cells is presumably induced by an excessive amount of BAFF. As a result of the above events, systemic inflammation and fibrosis are induced, and autoimmune-like manifestations are observed. This figure is created with BioRender.

proinflammatory and/or profibrotic phenotypes [35]. Consequently, inflammation and fibrosis are induced in cGVHD target organs [37]. The damaged thymic epithelial cells (required for the generation of Tregs as well as the negative selection) also cause a decrease in the number of Tregs [38]. Furthermore, the dysregulation of B cells causes autoreactive B cells to arise and produce autoreactive antibodies [39]. The emergence and activation of autoreactive B cells presumably stem from B cell exhaustion induced by aberrant levels of B cell-activating factor (BAFF) in the lymphoid microenvironment [40, 41].

4. Therapeutic targets and strategies for GVHD

4.1 TCR and BCR signaling

When the T cell receptor (TCR) interacts with an MHC-antigenic peptide complex, it induces molecular and cellular changes in T cells [42]. A wide range of signal transduction pathways in T cells is stimulated due to this interaction, leading to the activation of a variety of genes [43]. Effector enzymes such as kinases, phosphatases, and phospholipases are involved in the TCR signaling pathways, which are integrated by non-enzymatic adaptor proteins acting as a scaffold for interactions between proteins [42]. These intracellular signaling pathways can determine the features of immunity mediated by T cells [44].

The B cell receptor (BCR) complexes on inactivated B cells act as self-inhibiting oligomers [45]. The BCR signaling pathways are initiated, when BCR is bound to an antigen and induces actin-mediated nanoscale recombination of receptor clusters [46]. Due to this event, the BCR oligomers are opened and the ITAM domains are revealed, resulting in the transduction of intracellular signals which are crucial for B cell development, activation, proliferation, differentiation, and antibody production in health and disease [47].

In 2017, FDA approved ibrutinib, which targets B cells and T cells, for the treatment of cGVHD. Ibrutinib was the first FDA-approved drug for steroid-refractory cGVHD, and it was a significant milestone for GVHD research [48]. Ibrutinib is reported to modulate the functions of B cells and T cells by potently inhibiting Bruton's Tyrosine Kinase (BTK) and IL-2 Inducible T-cell Kinase (ITK) [49], which are involved in the B cell signaling and T cell signaling pathways, respectively. Treatment of cGVHD-affected recipients with ibrutinib resulted in decreased serum-autoantibodies and B-cell proliferation [50]. Data from the clinical trials show that symptoms of cGVHD improved in 67% of patients treated with ibrutinib [48].

4.2 Purinergic signaling

The Purinergic signaling pathways play a crucial role in a range of physiological systems including the immune system. In the purinergic signaling pathways, extracellular purine nucleosides and nucleotides such as adenosine and adenosine triphosphate (ATP) are used as signaling molecules that mediate the communication between cells through the activation of purinergic receptors [51]. There are four types of P1 (adenosine) receptors (A1, A2A, A2B, and A3). P2 receptors are subdivided into P2X and P2Y [52]. P2X receptors have seven subtypes (P2X1, P2X2, P2X3, P2X4, P2X5, P2X6, and P2X7), and P2Y receptors have 8 subtypes (P2Y1, P2Y2, P2Y4, P2Y6, P2Y11, P2Y12, P2Y13, and P2Y14) [52].

As demonstrated by several studies using mouse models of aGVHD, extracellular ATP is augmented in aGVHD-affected mice, and purinergic signaling is implicated in the pathogenic process of aGVHD (**Figure 3**) [53]. The conditioning regimens prior to allo-HSCT can induce tissue damage, leading to the release of DAMP molecules including ATP, which activates purinergic signaling [53]. The involvement of extracellular ATP is evidenced by the fact that the injection of the soluble ATP diphosphohydrolase (ATPDase) can reduce inflammation in aGVHD target organs and the serum level of IFNγ [53, 54].

Evidence suggests that; (i) P2X7 is a crucial P2X receptor in the development of aGVHD after the release of extracellular ATP, (ii) the expression of the P2X7 receptor is elevated in PBMCs in aGVHD patients, (iii) the liver, spleen, skin, and thymus in aGVHD-affected mice show the increased expression of the P2X7 receptor, (iv) the ATP-induced activation of the P2X7 receptor on host APCs can facilitate the stimulation, proliferation, and survival of donor T cells during aGVHD and (v) the P2X7 activation on host APCs may be associated with the expression of microRNA mir-155 [53, 55–57].

While the host P2X7 receptor is shown to play an integral role in the development of aGVHD, the donor P2X7 receptor is also a contributor to this disease. Evidence suggests that (i) the activation and proliferation of donor $CD4^+$ T cells and (ii) the metabolic fitness of donor $CD8^+$ T cells are also enhanced by the activated donor P2X7 receptor [58, 59]. In addition, the activation of P2X7 on donor Tregs can reduce their suppressive ability and stability of Tregs, promoting their conversion to Th17 cells [60].

Inhibition of the P2X7 receptor is reported to mitigate aGVHD in conventional and humanised mouse models of aGVHD. Treatment of allogeneic HSCT recipient mice with the P2X7 inhibitor pyridoxalphosphate-6-azophenyl-2′,4′-disulfonic acid (PPADS) can increase the survival rate and the number of Tregs, and reduce the

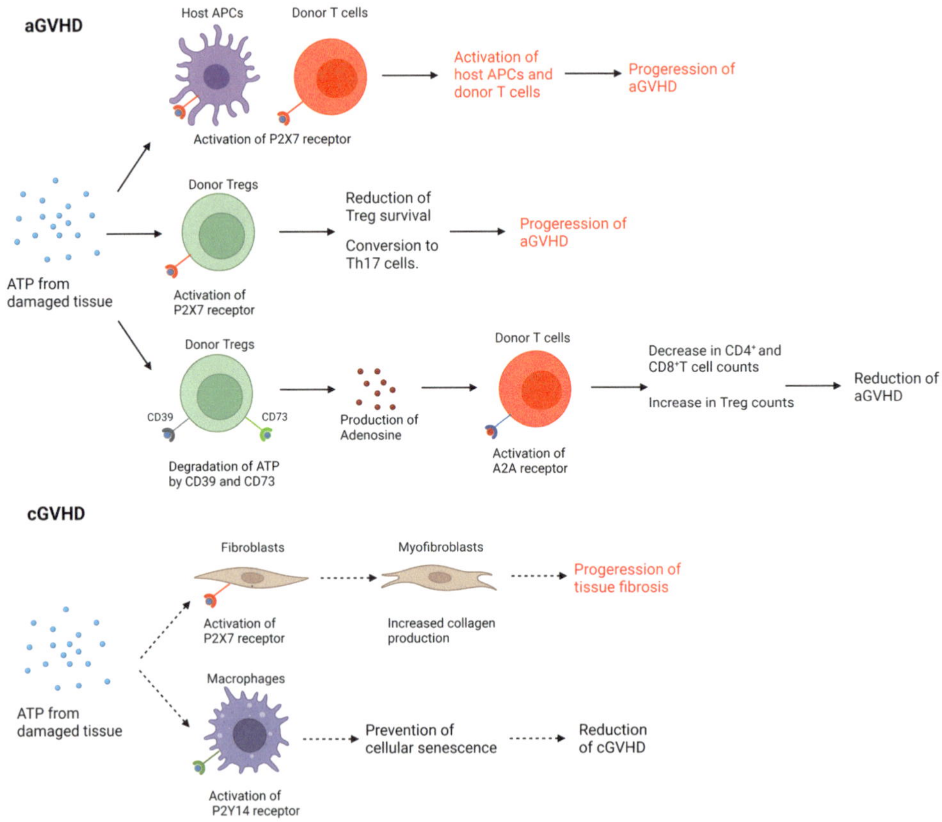

Figure 3.
Link between GVHD and the therapeutically targetable purinergic signaling pathways. In aGVHD, ATP is produced due to tissue damage. Host APCs and donor T cells can be activated by the P2X7 receptor, which results in the progression of aGVHD. The deactivation of donor Tregs can also be induced by the ATP-activated P2X7 receptor, which leads to the reduction of Treg survival and the progression of aGVHD. CD39 and CD73 on donor Tregs can degrade ATP to adenosine. Adenosine can activate the A2A receptor on donor T cells, which culminates in the decrease in the number of CD4⁺ and CD8⁺ T cells and the reduction of aGVHD. In cGVHD, ATP is also released because of tissue damage and may promote fibroblast-to-myofibroblast transition through the ATP-activated P2X7 receptor, leading to the augmented collagen production and the progression of tissue fibrosis. In contrast, the ATP-activated P2Y14 receiptor may prevent cellular senescence in macrophages and mitigate cGVHD. This figure is created with BioRender.

serum level of IFNγ and histological aGVHD [53, 54]. Administration of the P2X7 inhibitor brilliant blue G (BBG) to allogeneic HSCT recipient mice can also prevent weight loss and reduce inflammation in the liver and the production of inflammatory cytokines [56]. Furthermore, a crystal structure of the P2X7 receptor in complex with the inhibitor AZ10606120 has been reported (PDB: 5U1W) [61], and this structural information could be useful for the design and synthesis of novel P2X7 inhibitors which can be used in clinical settings.

The P2Y2 receptor is also reported to contribute to the pathogenesis of aGVHD [22, 57]. Evidence indicates that the number of cells expressing the P2Y2 receptor is increased in the intestinal tract in aGVHD-affected mice and that the increased P2Y2 expression enhances the severity of intestinal aGVHD [62]. Of note, knock-out allogeneic HSCT recipient mice of the P2Y2 receptor show an increased survival rate and decreased cytokine levels [62]. However, in the case where the P2Y2 receptor in donor cells is knocked out, no such improvement is observed [62]. In contrast, literature precedent suggests that the activation of the P2Y2 receptor can promote the migration of Tregs to sites of inflammation and thereby mitigate aGVHD [63]. Due to the dual functions of the P2Y2 receptor, targeting the P2Y2 receptor for the treatment has been challenging and there have been no reports about systemic injection of P2Y2 inhibitors/activators for the treatment of aGVHD [64].

While ATP is released in damaged tissues in allogeneic HSCT recipients and promotes inflammation, it is also degraded to adenosine by CD39 and CD73 [53]. In particular, a murine study indicates that CD39 and CD73 are highly expressed on $CD150^{high}$ Tregs [65]. As shown by a study using a mouse model of aGVHD, inhibition of CD39 and CD73 with adenosine 5′-(α,β-methylene)diphosphate (APCP) leads to the increase in the number of splenic $CD4^+$ and $CD8^+$ T cells, the serum levels of IFNγ and IL-6, and the mortality rate [66]. These data suggest that CD39 and CD73 play an alleviatory role in aGVHD. Evidence demonstrates that the production of adenosine by CD39 and CD73 results in the activation of the adenosine A2A receptor [66–68]. The activated A2A receptor can induce the expansion of donor Tregs and thereby mitigate aGVHD-induced inflammation [66–68]. The blockade of A2A with the antagonist SCH58261 exacerbates aGVHD by elevating the levels of TNFα, IFNγ, and IL-6 and the number of $CD4^+$ and $CD8^+$ T cells in sera [66]. In agreement with this report, the A2A agonist ATL-146e reduced weight loss and mortality in aGVHD-affected mice by (i) increasing serum IL-10 and reducing serum IFN-γ and IL-6, (ii) precluding the activation of splenic $CD4^+$ and $CD8^+$ T cells, and the infiltration of T cells into GVHD target organs [67]. Other A2A agonists, ATL-370 and ATL-1223, are reported to exert similar therapeutic effects on aGVHD [68]. Moreover, a crystal structure of the A2A receptor in complex with the activator ZM241385 has been reported (PDB: 5WF5) [69], and this structural information could facilitate the creation of novel A2A activators which can enter the clinic.

Although there are few to no reports about a link between purinergic signaling and cGVHD pathogenesis, activation of the P2X7 receptor is reported to promote fibroblast-to-myofibroblast transformation and contribute to the development of fibrosis [70]. The activation of the P2X7 receptor enhances Ca^{2+} influx and skews fibroblasts towards a fibrogenic phenotype, leading to augmented collagen production [70]. Considering fibrosis is a significant hallmark of cGVHD, the investigation into a correlation between purinergic signaling and fibroblast activity in cGVHD could open up a new window for the elucidation of mechanisms of cGVHD and the development of novel drugs for cGVHD (**Figure 3**). Furthermore, stress-induced cellular senescence in immune cells is reported to play a detrimental role in the pathogenesis of ocular cGVHD [71, 72], and a murine study indicates that the P2Y14 receptor modulates stress-induced cellular senescence in hematopoietic stem/progenitor cells [73]. Given these findings, the P2Y14 receptor may be a regulator of stress-induced cellular senescence in cGVHD, and development of agonists of the P2Y14 receptor could benefit cGVHD patients.

4.3 JAK/STAT signaling

The Janus kinase/signal transducer and activator of transcription (JAK/STAT) signaling pathways are regarded as a central communication junction for the immune system [74]. In the JAK/STAT signaling pathways, the cytoplasmatic kinase JAKs interact with the transcription factor STATs, and more than 50 cytokines and growth factors are involved in the JAK/STAT signaling pathways [75]. Mammals have 4 JAKs (JAK1, JAK2, JAK3, JAK4) and 7 STATs (STAT1, STAT2, STAT3, STAT4, STAT5a, STAT5b,

STAT6) [76], and the dysregulated JAK/STAT signaling pathways contribute to a variety of human diseases, which makes this signaling a promising drug target [77].

In the early phase of aGVHD, tissue damage due to the preconditioning regimen and the disease results in the release of DAMPs, leading to the increased expression of MHC on APCs at the infusion of donor cells [78]. Donor T cells are activated via direct or indirect allorecognition, and the activated donor T cells produce IFNγ to initiate the JAK/STAT signaling pathways through IFNγ receptors [78]. The resultant increase in the expression of the chemokine receptor CXCR3 on T cells enhances their migration to aGVHD target organs, which promotes tissue damage [79].

While clinical manifestations of cGVHD are different from those of aGVHD, they have similarities in some aspects of the pathogenic processes [80]. The JAK/STAT signaling pathways in the context of cGVHD have been intensively investigated [81]. Tregs play a crucial role in the reduction of cGVHD, and JAK1/JAK2 signaling pathways are thought to negatively regulate the development and proliferation of Tregs, as indicated by the fact that JAK2 inhibition can promote Treg proliferation [82, 83]. Tissue fibrosis is highly problematic in cGVHD, and M2-like macrophages producing TGF-β are presumably a key player [84]. IL-10 skews macrophages towards an M2-like phenotype through the IL-10 receptor-JAK1/STAT3 pathway [85]. Given these reports, it would be intriguing to investigate an association between macrophages and the JAK/STAT signaling pathways in the development of cGVHD-induced fibrosis.

Many researchers have focused on the development of inhibitors targeting JAK/STAT signaling pathways for the treatment of aGVHD and cGVHD [81]. As demonstrated by several preclinical data, inhibition of the JAK/STAT pathways can mitigate GVHD without affecting the GVL effect [81] Most recently, the JAK1/JAK2 inhibitor ruxolitinib has been approved by FDA for aGVHD and cGVHD. In 2019, FDA approved ruxolitinib to treat steroid-refractory aGVHD patients 12 years or older [86]. The clinical trials show that the day-28 overall response rate (ORR) was 100% for Grade 2 aGVHD, 40.7% for Grade 3 aGVHD, and 44.4% for Grade 4 aGVHD [86]. In 2021, FDA approval was also granted to ruxolitinib for the therapy of steroid-resistant cGVHD patients 12 years or older [87]. The clinical trial data demonstrate that the ORR was 70%, and the median durations of response, which were calculated from first response to progression, death, or new systemic therapies for cGVHD, were 4.2 months [87]. A crystal structure of JAK2 in complex with ruxolitinib is provided in the PDB database (PDB: 6VGL) [88], and this structural information could be useful for the design of more potent and selective JAK1/JAK2 inhibitors. Another promising JAK1 inhibitor is itacitinib [89]. Data from a phase 1 clinical trial of itacitinib shows that 70.6% of steroid-refractory cGVHD patients were treated in a satisfactory manner [90]. Furthermore, two clinical trials of itacitinib for cGVHD have recently commenced (ClinicalTrials.gov identifier: NCT04200365, NCT03584516). It is of great medical significance that novel drugs targeting the JAK/STAT signaling will continue to be developed for the treatment of aGVHD and cGVHD.

4.4 NF-κB signaling

The transcription factor nuclear factor kappa B (NF-κB) controls the expression of various genes important for the induction of inflammatory responses in innate and adaptive immune cells [91]. NF-κB is a family of heterodimers or homodimers generated from different combinations of the following 5 proteins: p65/RelA, RelB, c-Rel, p105/p50 (NF-κB1), and p100/p52 (NF-κB2) [92]. Among them, the p50/p65 complex is thought to be the most abundant form of NF-κB dimer [93]. When NF-κB is inactive, it is retained in the cytoplasm by the IκB family of inhibitors [94, 95]. In response to a wide range of stimuli such as the proinflammatory cytokines IL-1 and TNF-α, IκB kinase (IKK) is activated to phosphorylate the 2 serine residues of IκBα [96]. The phosphorylation causes the 26S proteasome to induce the ubiquitination and degradation of IKβ. Subsequently, NF-κB is translocated into the nucleus and triggers gene transcription, leading to the production of proteins necessary for immune responses [97]. Thus, NF-κB is regarded as a therapeutic target for the treatment of various inflammatory diseases.

The NF-κB signaling pathways have captured increasing attention from GVHD researchers. It has been reported that the activation of RelB in APCs contributes to the expansion of donor Th1 cells and subsequent alloreactivity, which leads to the development of aGVHD [98]. The NF-kB signaling pathways can be survival and proliferation signals and contribute to B-cell alloantibody deposition and germinal center formation, which play a critical role in the pathogenic process of cGVHD [99, 100].

Bortezomib is an FDA-approved drug for the treatment of multiple myeloma and is known to be an indirect inhibitor of NF-κB [101]. A murine study suggests that aGVHD can be prevented by treatment with bortezomib early after allogeneic HSCT [102, 103]. Bortezomib is undergoing clinical trials for aGVHD (BMT CTN 1203), and the phase1/2 study shows that bortezomib can be used in combination with tacrolimus and methotrexate in a tolerable immunosuppressive regimen after allogeneic HSCT [104]. Bortezomib can also be effective for the treatment of cGVHD. NF-κB inhibition with Bortezomib is suggested to cause apoptosis of germinal center B cells during reconstitution, leading to the decrease in donor-derived B cell numbers and BAFF expression [103]. With these promising data, clinical trials of bortezomib for the treatment of steroid-refractory cGVHD are in progress (NCT01158105). At present, there are no NF-κB inhibitors approved by FDA for aGVHD or cGVHD. Generally, direct inhibitors are superior to indirect ones in terms of selectivity. Thus, novel direct NF-κB inhibitors with high selectivity are greatly anticipated for the treatment of GVHD.

4.5 Hedgehog signaling

The Hedgehog signaling pathways are involved in the regulation of cell proliferation, survival, and differentiation [105], and its aberrant activation contributes to detrimental events such as the self-renewal and metastasis of cancer stem cells [106]. In the absence of Hedgehog ligand (Hh), the activation of Smoothened (SMO) is inhibited by Patched (PTCH) [107]. Subsequently, the activity of glioma-associated oncogene homolog (Gli) is suppressed by a protein complex mainly composed of a suppressor of fused (SUFU), which phosphorylates Gli and prevents it from entering the nucleus. In the presence of Hh, the binding of Hh to PTCH precludes the SMO inhibition mediated by PTCH [107]. Activated SMO prevents phosphorylation of Gli mediated by the SUFU complex, leading to the migration of Gli to the nucleus and the induction of downstream target gene expression [107].

Fibrosis is a highly problematic feature of cGVHD, and a profibrotic activity of Hedgehog signaling in patients and mouse models of cGVHD has been reported [108]. Overexpression of Hh, which is an inducer of the Hedgehog signaling pathways, is observed in human and murine sclerodermatous cGVHD [108]. The downstream processes of the Hedgehog signaling pathway cause overexpression of Gli-1 and Gli-2, particularly in fibroblasts [109]. The abnormal expression of Gli-1 and Gli-2 may result in the overproduction of collagen and the resultant pathologic fibrosis in cGVHD target organs [109]. Furthermore, the Hedgehog signaling is suggested to contribute to the increase of profibrotic M2-like macrophages in the cGVHD-affected skin [109].

There are several inhibitors of the Hedgehog pathways. Among others, sonidegib, vismodegib, and glasdegib are SMO inhibitors approved by FDA for the treatment of basal cell carcinoma [110]. These 3 SMO inhibitors are currently undergoing clinical trials for cGVHD therapy (NCT02086513, NCT02337517, NCT04111497). According to a report of the Phase-1 trial of sonidegib, where 17 steroid-refractory cGVHD patients participated, protein expression of hedgehog signaling pathway molecules was decreased by treatment with sonidegib as judged by immunohistochemical evaluation of the skin [111]. With respect to the creation of novel SMO inhibitors for the treatment of GVHD, Lacroix et al. found a potential SMO inhibitor by performing structure-based virtual screening of 3.2 million available, lead-like molecules against SMO and subsequent biological validations of the top-ranked compounds [112]. This information could benefit the design and synthesis of more potent and selective inhibitors of SMO.

4.6 Endoplasmic reticulum stress

While elucidation of mechanisms of cGVHD is still elusive, chronic inflammation is characteristic of cGVHD [113]. Senescent macrophages contribute to ocular cGVHD in mice, and gray eyebrows, skin wrinkles and conjunctival cancer are observed in human cGVHD [71, 114]. These findings suggest that ageing in donor- and recipient-derived cells is induced in cGVHD [71]. Evidence suggests that chronic inflammation and age-related diseases are associated with the elevation of endoplasmic reticulum (ER) stress [115, 116]. Mukai et al found that ER stress was increased in organs affected by cGVHD in mice [117]. Treatment of cGVHD-affected mice with the known ER stress reducer 4-phenylburyric acid (PBA) resulted in mitigation of systemic inflammation and fibrosis induced by cGVHD [117]. Of note, PBA is approved by FDA for the treatment of urea cycle disorders, and its safety was proven [118]. Investigation at the cellular level indicates that ER stress contributes to fibrosis as well as inflammation induced by cGVHD. Elevated ER stress caused (i) the dysregulation of lacrimal-gland-derived fibroblasts and (ii) abnormal production of MCP-1/CCL2, IL-6, and connective tissue growth factor (CTGF) [117]. Suppression of ER stress with PBA reduced their abnormal production of the inflammatory and fibrotic molecules [117]. In addition, ER stress induced by cGVHD skewed splenic macrophages towards an M2-like phenotype, and treatment of them with PBA promoted their differentiation into an M1-like phenotype [117]. Several reports also indicate that the augmentation of M2-like macrophages is implicated in the progression of cGVHD [84, 119, 120]. M2-like macrophages are thought to contribute to the pathogenesis of fibrosis-associated diseases [121], and it seems to be the case with cGVHD. As these analyses were performed in a bulk population, further investigation will be needed. Macrophages and fibroblasts are known to be heterogeneous populations [122–125]. In particular, mounting evidence suggests that macrophage heterogeneity is multidimensional and more complex than M1/M2 classification [126]. Hence, single-cell analyses could greatly facilitate the understanding of a correlation between ER stress and macrophages/fibroblasts in the development of cGVHD and make ER stress a more compelling therapeutic target for cGVHD therapy.

4.7 Aberrant immune cell infiltration

While aGVHD and cGVHD show different clinical manifestations, one of their common features is abnormal immune cell infiltration, which results in organ damage and severe inflammation and fibrosis. Mukai et al devised a novel therapeutic strategy for both types of GVHD by targeting vascular adhesion protein-1 (VAP-1) [127], which is known to be overexpressed in inflamed organs [128]. VAP-1 is an endothelial surface glycoprotein assisting leucocyte migration from the bloodstream to tissues and possesses the following 2 functional domains: a distal adhesion domain and a catalytic amine oxidase domain [129]. For infiltration into tissues, the amino group in leukocytes undergoes a nucleophilic attack on the carbonyl group in VAP-1 [129]. The subsequent catalytic conversion of the primary amine to the corresponding aldehyde allows immune cells to squeeze into tissues through blood vessels [129, 130]. Pursuant to their study with the use of a mouse model where aGVHD shifts to cGVHD [127], (i) the protein expression of VAP-1 is increased in organs with GVHD, where the number of inflammatory cells is accordingly augmented, (ii) blockade of VAP-1 with a novel inhibitor reduced the number of tissue-infiltrating leukocytes and thereby mitigated GVHD manifestations such as inflammation and fibrosis and (iii) the VAP-1 inhibition caused few to no severe adverse effects. Collectively, inhibition of VAP-1 could be an effective all-in-one approach for the treatment of aGVHD and cGVHD.

4.8 NOTCH signaling

The Notch signaling pathways are cell-to-cell communication induced by interactions between Notch receptors (NOTCH1, NOTCH2, NOTCH3, and NOTCH4) and NOTCH ligands (Jagged1 (JAG1), JAG2, Delta-like 1 (DLL1), DLL3 and DLL4) [131]. Due to these intercellular interactions, the NOTCH receptor is proteolytically activated by an ADAM family metalloprotease and subsequently by the γ-secretase complex [132]. The sequential cleavages lead to the release of the intracellular NOTCH domain (NICD), which is a transcriptionally active fragment [133]. NICD migrates to the nucleus and binds to the DNA binding CSL/RBP-Jk factor, forming a transcriptional activation complex with a mastermind-like (MAML) family coactivator [133]. This final complex triggers the transcription of target genes which are important for biological processes such as proliferation, differentiation, and survival [134].

A correlation between the Notch signaling pathways and alloimmune responses has gained interest from GVHD researchers. Studies using animal models of aGVHD suggest that; (i) the Notch signaling promotes activation, differentiation, and alloreactivity of T cells [135] and (ii) dendritic cells with high DLL4 expression show an increase in the production of IFN-γ and IL-17 [136]. The Notch signaling is also implicated in the pathogenic process of cGVHD. A murine study shows that NOTCH1 and NOTCH2 as well as DLL1 and DLL4 serve significant functions in regulating proinflammatory cytokine production by T cells [137]. Investigation using *in-vitro* human B-cell assay systems demonstrates that abnormal activation of NOTCH2 is correlated with hyperresponsiveness of BCR on B cells from cGVHD patients [138].

GVHD treatments by targeting the Notch signaling pathway have been reported. A series of experiments using a mouse model of aGVHD reveals; (i) inhibitors of γ-secretase block proteolytic activation of all the NOTCH receptors, but has severe toxicity in the gut epithelium, (ii) NOTCH1 inhibition using an antibody mitigates GVHD but causes serious toxicity and (iii) treatment with a combination of anti-DLL1 and anti-DLL4 reduces aGVHD without debilitating adverse effects while maintaining a GVL effect of donor T cells [139]. An anti-DLL1 antibody is also effective for the treatment of murine cGVHD in combination with an anti-DLL4 antibody [137]. Treatment with all-trans-retinoic acid (ATRA) prevents NOTCH2-induced BCR hyperresponsiveness, which plays a detrimental role in cGVHD pathogenesis [137]. It appears that NOTCH2 and DLL1/4 are promising drug targets for the treatment of the 2 types of GVHD. Therefore, it is highly anticipated that novel, selective inhibitors of NOTCH2 and DLL1/4 will be developed for use in human GVHD.

4.9 Rho/ROCK signaling

Rho-associated coiled-coil-containing protein kinases (ROCKs) are serine-threonine-specific protein kinases, and mammals have ROCK1 and ROCK2 [140]. ROCKs are downstream effector proteins of GTPase Rho, and abnormal activation of the Rho/ROCK pathways contributes to the development of various diseases [140]. In particular, ROCK2 is known to regulate (i) the balance of Th17 cells and Tregs and (ii) profibrotic pathways [141]. ROCK2 activation increases Th17 cell-specific transcription factors by promoting STAT3 phosphorylation [142]. In addition, when ROCK2 is activated by profibrotic mediators such as tumor growth factor-β (TGF-β), it causes myocardin-related transcription factors to activate profibrotic genes in fibroblasts [143, 144]. This profibrotic gene activation induces fibroblast-to-myofibroblast differentiation and the resultant increase in collagen production [143, 144].

A study using a cGVHD mouse model shows that treatment with belumosudil, which is a selective ROCK2 inhibitor, can substantially reduce cGVHD-induced fibrosis in the lung [145]. In 2021, belumosudil was approved by FDA for the treatment of cGVHD, and the clinical trial data show that the overall response rate was 75% (6% complete response and 69% partial response) [146].

ROCK1 is also thought to be involved in the development of fibrosis, and pan-ROCK inhibitors targeting ROCK1/2 are thereby expected to show better treatment outcomes for cGVHD [147]. Several pan-ROCK inhibitors have been granted approval for human use [148–151] In particular, netarsudil has been approved by FDA for the treatment of glaucoma [151]. However, due to a lack of overall kinome selectivity of the reported dual ROCK1/2 inhibitors, there is still scope for improvement in pan-ROCK inhibitors [152]. Hu et al. has recently reported the synthesis and *in-vitro* evaluation of a novel series of 5*H*-chromeno[3,4-c]pyridine, 6*H*-isochromeno[3,4-c]pyridine, and 6*H*-isochromeno[4,3-d]pyrimidine derivatives as dual ROCK1/2 inhibitors [152]. Their data show that some of the novel pan-ROCK inhibitors display potent inhibitory activity against ROCK1/2 and possess excellent kinome selectivity [152]. They also provided a crystal structure of ROCK2 in complex with one of the novel dual ROCK1/2 inhibitors (PDB ID: 7JNT). This structural information can be useful in the structure-based design of other new pan-ROCK inhibitors.

5. Conclusion

While recent decades have seen significant technological and medical advances, aGVHD and cGVHD are still a major hurdle to successful allogeneic HSCT in clinical settings. Systemic corticosteroid therapy, with or without immunosuppressive agents, is the first-line treatment for GVHD, although it can cause severe adverse effects and approximately 50% of GVHD patients develop steroid-resistant GVHD. Thus, sophisticated treatments of steroid-refractory aGVHD and cGVHD are highly anticipated by medical settings. A great deal of effort has been invested in the elucidation of mechanisms of GVHD and development of safe and efficacious drugs for GVHD. Recently, several drugs have been approved by FDA for the treatment of steroid-refractory aGVHD and cGVHD. Despite this progress, there is still a need to create novel drugs with better efficacy for GVHD therapy. This chapter focused on druggable targets for the treatment of GVHD with an aim to stimulate various GVHD researchers (from medicinal chemists to biologists) to create novel drugs which can enter the clinic. While several signaling pathways have been intensively studied in the context of GVHD, there are underexplored signaling pathways. In particular, the purinergic signaling pathway is one of the understudied signaling pathways in GVHD. The P2X7, A2A, and P2Y14 receptors seem to be compelling drug targets for the treatment of GVHD, and clinical settings could benefit from safe and efficacious (i) inhibitors of the P2X7 receptor and (ii) activators of the A2A and/or P2Y14 receptors. However, the development of new drugs is a costly and time-consuming process. To overcome this setback, the use of AL/ML has captured great interest from many researchers and has been expected to substantially reduce the cost and time of drug development. A combination of AL/ML and molecular design could greatly facilitate the development of novel, effective, safe, and affordable drugs for the treatment of GVHD.

Author details

Shin Mukai
Beneras Pharma Inc., Boston, United States of America

*Address all correspondence to: shin.mukai@beneras-pharma.com

References

[1] Jagasia M, Arora M, Flowers MED, Chao NJ, McCarthy PL, Cutler CS, et al. Risk factors for acute GVHD and survival after hematopoietic cell transplantation. Blood. 2012;**119**:296-307. DOI: 10.1182/blood-2011-06-364265

[2] Vigorito AC, Campregher PV, Storer BE, Carpenter PA, Moravec CK, Kiem H-P, et al. Evaluation of NIH consensus criteria for classification of late acute and chronic GVHD. Blood. 2009;**114**:702-708. DOI: 10.1182/blood-2009-03-208983

[3] Malard F, Huang X-J, Sim JPY. Treatment and unmet needs in steroid-refractory acute graft-versus-host disease. Leukemia. 2020;**34**:1229-1240

[4] Ceredig R. Graft-versus-host disease: Who's responsible? Immunology and Cell Biology. 2012;**90**:253-254. DOI: 10.1038/icb.2011.62

[5] Martin PJ, Levine DM, Storer BE, Warren EH, Zheng X, Nelson SC, et al. Genome-wide minor histocompatibility matching as related to the risk of graft-versus-host disease. Blood. 2017;**129**:791-798. DOI: 10.1182/blood-2016-09-737700

[6] Lee JW. Prevention of chronic GVHD. Best Practice & Research Clinical Haematology. 2008;**21**:259-270. DOI: 10.1016/j.beha.2008.02.010

[7] Wolff D, Fatobene G, Rocha V, Kröger N, Flowers ME. Steroid-refractory chronic graft-versus-host disease: Treatment options and patient management. Bone Marrow Transplantation. 2021;**56**:2079-2087

[8] Sung AD, Chao NJ. Concise review: Acute graft-versus-host disease: Immunobiology, prevention, and treatment. Stem Cells Translational Medicine. 2013;**2**:25-32. DOI: 10.5966/sctm.2012-0115

[9] Nassereddine S, Rafei H, Elbahesh E, Tabbara I. Acute graft versus host disease: A comprehensive review. Anticancer Research. 2017;**37**:1547-1555

[10] Teshima T, Reddy P, Zeiser R. Acute graft-versus-host disease: Novel biological insights. Biology of Blood and Marrow Transplantation. 2016;**22**:11-16

[11] Ballester-Sánchez R, Navarro-Mira M, Sanz-Caballer J, Botella-Estrada R. Review of cutaneous graft-vs-host disease. Actas Dermo-Sifiliográficas. 2016;**107**:183-193

[12] Zeiser R, Teshima T. Nonclassical manifestations of acute GVHD. Blood. 2021;**138**:2165-2172. DOI: 10.1182/blood.2021012431

[13] Lee SJ. Classification systems for chronic graft-versus-host disease (de novo cGVHD). Blood. 2017;**129**:30-37. DOI: 10.1182/blood-2016-07686642

[14] Ratanatharathorn V, Ayash L, Lazarus HM, Fu J, Uberti JP. Chronic graft-versus-host disease: Clinical manifestation and therapy. Bone Marrow Transplantation. 2001;**28**:121-129. DOI: 10.1038/sj.bmt.1703111

[15] Saidu NEB, Bonini C, Dickinson A, Grce M, Inngjerdingen M, Koehl U, et al. New approaches for the treatment of chronic graft-versus-host disease: Current status and future directions. Frontiers in Immunology. 2020;**11**:578314. DOI: 10.3389/fimmu.2020.578314

[16] Filipovich AH, Weisdorf D, Pavletic S, Socie G, Wingard JR, Lee SJ, et al. National

Institutes of Health consensus development project on criteria for clinical trials in chronic graft-versus-host disease: I. Diagnosis and staging working group report. Biology of Blood and Marrow Transplantation. 2005;**11**:945-955

[17] Pavletic SZ, Vogelsang GB, Lee SJ. National Institutes of Health Consensus Development Project on Criteria for Clinical Trials in Chronic Graft-versus-Host Disease: Preface to the series. Biology of Blood and Marrow Transplantation. 2014;**2014**(21):387-388. DOI: 10.1016/j.bbmt.2014.12.035

[18] Sarantopoulos S, Cardones AR, Sullivan KM. How I treat refractory chronic graft-versus-host disease. Blood. 2019;**133**:1191-1200. DOI: 10.1182/blood-2018-04-785899

[19] Rashidi A, DeFor TE, Holtan SG, Blazar BR, Weisdorf DJ, MacMillan ML. Outcomes and predictors of response in steroid-refractory acute graft-versus-host disease. Biology of Blood and Marrow Transplantation. 2019;**25**:2297-2302. DOI: 10.1016/j.bbmt.2019.07.017

[20] What to know about corticosteroids. Available from: https://www.medicalnewstoday.com/articles/corticosteroids

[21] Choi SW, Levine JE, Ferrara JLM. Pathogenesis and management of graft versus host disease. Immunology and Allergy Clinics of North America. 2010;**30**:75-101. DOI: 10.1016/j.iac.2009.10.001

[22] Apostolova P, Zeiser R. The role of purine metabolites as DAMPs in acute graft-versus-host disease. Frontiers in Immunology. 2016;7:439

[23] Mosmann TR, Cherwinski H, Bond MW, Giedlin MA, Coffman RL. Pillars article: Two types of Murine Helper T cell clone. I. Definition according to profiles of lymphokine activities and secreted proteins. Journal of Immunology. 1986;**136**:2348-2357

[24] Park H, Li Z, Yang XO, Chang SH, Nurieva R, Wang Y-H, et al. A distinct lineage of CD4 T cells regulates tissue inflammation by producing interleukin 17. Nature Immunology. 2005;**6**:1133-1141. DOI: 10.1038/ni1261

[25] Ho VT, Soiffer RJ. The history and future of T-cell depletion as graft-versus-host disease prophylaxis for allogeneic hematopoietic stem cell transplantation. Blood. 1998;**98**:3192-3204

[26] Reddy P, Ferrara JLM. Immunobiology of acute graft-versus-host disease. Blood Reviews. 2003;**17**:187-194. DOI: 10.1016/s0268-960x(03)00009-2

[27] Schwab L, Goroncy L, Palaniyandi S, Gautam S, Triantafyllopoulou A, Mocsai A, et al. Neutrophil granulocytes recruited upon translocation of intestinal bacteria enhance graft-versus-host disease via tissue damage. Nature Medicine. 2014;**20**:648-654

[28] Hülsdünker J, Ottmüller KJ, Neeff HP, Koyama M, Gao Z, Thomas OS, et al. Neutrophils provide cellular communication between ileum and mesenteric lymph nodes at graft-versus-host disease onset. Blood. 2018;**131**:1858-1869. DOI: 10.1182/blood-2017-10-812891

[29] Jardine L, Cytlak U, Gunawan M, Reynolds G, Green K, Wang X-N, et al. Donor monocyte-derived macrophages promote human acute graft-versus-host disease. Journal of Clinical Investigation. 2020;**130**:4574-4586. DOI: 10.1172/JCI133909

[30] Terakura S, Martin PJ, Shulman HM, Storer BE. Cutaneous macrophage

infiltration in acute GvHD. Bone Marrow Transplantation. 2015;**50**:1135-1137. DOI: 10.1038/ bmt.2015.114

[31] Ianni MD, Falzetti F, Carotti A, Terenzi A, Castellino F, Bonifacio E, et al. Tregs prevent GVHD and promote immune reconstitution in HLA-haploidentical transplantation. Blood. 2011;**117**:3921-3928. DOI: 10.1182/ blood-2010-10311894

[32] Guo W-w, Su X-h, Wang M-y, Han M-z, Feng X-m, Jiang E-l. Regulatory T cells in GVHD therapy. Frontiers in Immunology. 2021;**12**:9676

[33] Olson JA, Leveson-Gower DB, Gill S, Baker J, Beilhack A, Negrin RS. NK cells mediate reduction ofGVHD by inhibiting activated, alloreactive T cells while retaining GVT effects. Blood. 2010;**115**:4293-4301. DOI: 10.1182/ blood-2009-05-222190

[34] Ullrich E, Salzmann-Manrique E, Bakhtiar S, Bremm M, Gerstner S, Herrmann E, et al. Relation between acute GVHD and NK cell subset reconstitution following allogeneic stem cell transplantation. Frontiers in Immunology. 2016;7:595. DOI: 10.3389/ fimmu.2016.00595

[35] Zeiser R, Blazar BR. Pathophysiology of chronic graft-versus-host disease and therapeutic targets. New England Journal of Medicine. 2017;**377**:2565-2579. DOI: 10.1056/NEJMra1703472

[36] Wu T, Young JS, Johnston H, Ni X, Deng R, Racine J, et al. Thymic damage, impaired negative selection, and development of chronic graft-versus-host disease caused by donor $CD4^{+}$ and $CD8^{+}$ T cells. Journal of Immunology. 2013;**191**:488-499. DOI: 10.4049/ jimmunol.1300657

[37] Blazar BR, Murphy WJ, Abedi M. Advances in graft-versus-host disease biology and therapy. Nature Reviews Immunology. 2012;**12**:443-458. DOI: 10.1038/nri3212

[38] Zorn E, Kim HT, Lee SJ, Floyd BH, Litsa D, Arumugarajah S, et al. Reduced frequency ofFOXP3⩽ CD4⩽CD25⩽ regulatory T cells in patients with chronic graft-versus-host disease. Blood. 2005;**106**:2903-2911. DOI: 10.1182/ blood-2005-03-1257

[39] Kapur R, Ebeling S, Hagenbeek A. B-cell involvement in chronic graft-versus-host disease. Haematologica. 2008;**93**:1702-1717. DOI: 10.3324/ haematol.13311

[40] Pers J-O, Daridon C, Devauchelle V, Jousse S, Saraux A, Jamin C, et al. BAFF overexpression is associated with autoantibody production in autoimmune diseases. Annals of the New York Academy of Sciences. 2005;**1050**:34-39. DOI: 10.1196/annals.1313.004

[41] Khoder A, Alsuliman A, Basar R, Sobieski C, Kondo K, Alousi AM, et al. Evidence for B cell exhaustion in chronic graft-versus-host disease. Frontiers in Immunology. 2018;**8**:1937. DOI: 10.3389/ fimmu.2017.01937

[42] Samelson LE. Signal transduction mediated by the T cell antigen receptor: The role of adapter proteins. Annual Review of Immunology. 2002;**20**:371-394. DOI: 10.1146/annurev. immunol.20.092601.11135

[43] Cantrell DA. T-cell antigen receptor signal transduction. Immunology. 2002;**105**:369-374. DOI: 10.1046/j.1365-2567.2002.01391.x

[44] Hwang J-R, Byeon Y, Kim D, Park S-G. Recent insights of T cell receptor-mediated signaling pathways for T cell activation and development. Experimental & Molecular Medicine.

2020;**52**:750-761. DOI: 10.1038/s12276-020-0435-8

[45] Wena Y, Jinga Y, Yanga L, Kanga D, Jianga P, Lib N, et al. The regulators of BCR signaling during B cell activation. Blood Science. 2019;**1**:119-129. DOI: 10.1097/BS9.0000000000000026

[46] Fleire SJ, Goldman JP, Carrasco YR, Weber M, Bray D, Batista FD. B cell ligand discrimination through a spreading and contraction response. Science. 2006;**312**:738-741. DOI: 10.1126/science.1123940

[47] Tsourkas PK, Das SC, Yu-Yang P, Liu W, Pierce SK, Raychaudhuri S. Formation of BCR oligomers provides a mechanism for B cell affinity discrimination. Journal of Theoretical Biology. 2012;**307**:174-182. DOI: 10.1016/j.jtbi.2012.05.008

[48] FDA approves treatment for chronic graft versus host disease. 2017 Available from: https://www.fda.gov/news-events/press-announcements/fda-approves-treatment-chronic-graft-versus-host-disease

[49] Miklos D, Cutler CS, Arora M, Waller EK, Jagasia M, Pusic I, et al. Ibrutinib for chronic graft-versus-host disease after failure of prior therapy. Blood. 2017;**130**:2243-2250. DOI: 10.1182/blood-2017-07-793786

[50] Schutt SD, Fu J, Nguyen H, Bastian D, Heinrichs J, Wu Y, et al. Inhibition of BTK and ITK with ibrutinib is effective in the prevention of chronic graft-versus-host disease in mice. PLoS One. 2015;**10**:e0137641. DOI: 10.1371/journal.pone.0137641

[51] Praetorius HA, Leipziger J. Intrarenal purinergic signaling in the control of renal tubular transport. Annual Review of Physiology. 2009;**72**:377-393

[52] Kong Q, Quan Y, Tian G, Zhou J, Liu X. Purinergic P2 receptors novel mediators of mechanotransduction. Frontiers in Pharmacology. 2021;**12**: 671809

[53] Wilhelm K, Ganesan J, Müller T, Dürr C, Grimm M, Beilhack A, et al. Graft-versus-host disease is enhanced by extracellular ATP activating P2X7R. Nature Medicine. 2010;**16**:1434-1438

[54] Koehn BH, Saha A, McDonald-Hyman C, Loschi M, Thangavelu G, Ma L, et al. Danger-associated extracellular ATP counters MDSC therapeutic efficacy in acute GVHD. Blood. 2019;**134**:1670-1682. DOI: 10.1182/blood.2019001950

[55] Zhong X, Zhu F, Qiao J, Zhao K, Zhu S, Zeng L, et al. The impact of P2X7 receptor antagonist, brilliant blue G on graft-versus-host disease in mice after allogeneic hematopoietic stem cell transplantation. Cellular Immunology. 2016;**310**:71-77. DOI: 10.1016/j.cellimm.2016.07.014

[56] Cuthbertson P, Adhikary SR, Geraghty NJ, Guy TV, Hadjiashrafi A, Fuller SJ, et al. Increased P2X7 expression in the gastrointestinal tract and skin in a humanised mouse model of graft-versus-host disease. Clinical Science. 2020;**134**:207-223. DOI: 10.1042/CS20191086

[57] Chen S, Smith BAH, Iype J, Prestipino A, Pfeifer D, Grundmann S, et al. MicroRNA-155-deficient dendritic cells cause less severe GVHD through reduced migration and defective inflammasome activation. Blood. 2015;**2015**:103-112. DOI: 10.1182/blood-2014-12-617258

[58] Yip L, Woehrle T, Corriden R, Hirsh M, Chen Y, Inoue Y, et al. Autocrine regulation of T-cell activation by

ATP release and P2X7 receptors. FASEB Journal. 2009;**23**:1685-1693. DOI: 10.1096/fj.08-126458

[59] Silva H, Beura LK, Wang H, Hanse EA, Gore R, Scott MC, et al. The purinergic receptor P2RX7 directs metabolic fitness of long-lived memory CD8 + T cells. Nature. 2018;**559**:264-268. DOI: 10.1038/s41586-018-0282-0

[60] Schenk U, Frascoli M, Proietti M, Geffers R, Traggiai E, Buer J, et al. ATP inhibits the generation and function of regulatory T cells through the activation of purinergic P2X receptors. Science Signaling. 2011;**4**:ra12

[61] Karasawa A, Kawate T. Structural basis for subtype-specific inhibition of the P2X7 receptor. eLife. 2016;**5**:e22153. DOI: 10.7554/eLife.22153

[62] Klämbt V, Wohlfeil SA, Schwab L, Hülsdünker J, Ayata K, Apostolova P, et al. A novel function for P2Y2 in myeloid recipient-derived cells during graft-versus-host disease. Journal of Immunology. 2015;**195**:5795-5804. DOI: 10.4049/jimmunol.1501357

[63] Dürr C, Follo M, Idzko M, Reichardt W, Zeiser R. Graft-versus-host disease reduces regulatory T-cell migration into the tumour tissue. Immunology. 2012;**137**:80-88. DOI: 10.1111/j.1365-2567.2012.03610.x

[64] Cuthbertson P, Geraghty NJ, Adhikary SR, Bird KM, Fuller SJ, Watson D, et al. Purinergic signalling in allogeneic haematopoietic stem cell transplantation and graft-versus-host disease. International Journal of Molecular Sciences. 2021;**22**:8343. DOI: 10.3390/ijms22158343

[65] Yang S, Sheng X, Xiang D, Wei X, Chen T, Yang Z, et al. CD150 high treg cells may attenuate graft versus host disease and intestinal cell apoptosis after hematopoietic stem cell transplantation. American Journal of Translational Research. 2019;**11**:1299-1310

[66] Tsukamoto H, Chernogorova P, Ayata K, Gerlach UV, Rughani A, Ritchey JW, et al. Deficiency of CD73/ecto-5'-nucleotidase in mice enhances acute graft-versus-host disease. Blood. 2012;**119**:4554-4564. DOI: 10.1182/blood-2011-09-375899

[67] Lappas CM, Liu P-C, Linden J, Kang EM, Malech HL. Adenosine A2A receptor activation limits graft-versus-host disease after allogenic hematopoietic stem cell transplantation. Journal of Leukocyte Biology. 2009;**87**:354

[68] Han KL, Thomas SVM, Koontz SM, Changpriroa CM, Ha S-K, Malech HL, et al. Adenosine A_2A receptor agonist-mediated increase in donor-derived regulatory T cells suppresses development of graft-versus-host disease. Journal of Immunology. 2012;**190**:458-468. DOI: 10.4049/jimmunol.1201325

[69] White KL, Eddy MT, Gao Z-G, Han GW, Lian T, Deary A, et al. Structural connection between activation microswitch and allosteric sodium site in GPCR signaling. Structure. 2018;**26**:259-269. DOI: 10.1016/j.str.2017.12.013

[70] Gentile D, Lazzerini PE, Gamberucci A, Natale M, Selvi E, Vanni F, et al. Searching novel therapeutic targets for scleroderma: P2X7-receptor is up-regulated and promotes a fibrogenic phenotype in systemic sclerosis fibroblasts. Frontiers in Pharmacology. 2017;**8**:638

[71] Kawai M, Ogawa Y, Shimmura S, Ohta S, Suzuki T, Kawamura N, et al. Expression and localization of aging markers in lacrimal gland of chronic graft-versus-host disease. Scientific

Reports. 2013;**3**:2455. DOI: 10.1038/srep02455

[72] Yamane M, Sato S, Shimizu E, Shibata S, Hayano M, Yaguchi T, et al. Senescence-associated secretory phenotype promotes chronic ocular graft-vs-host disease in mice and humans. FASEB Journal. 2020;**34**:10778-10800. DOI: 10.1096/fj.201900218R

[73] Cho J, Yusuf R, Kook S, Attar E, Lee D, Park B, et al. Purinergic P2Y14 receptor modulates stress-induced hematopoietic stem/progenitor cell senescence. Journal of Clinical Investigation. 2014;**124**:3159-3171. DOI: 10.1172/JCI61636

[74] Villarino AV, Kanno Y, O'Shea JJ. Mechanisms and consequences of Jak-STAT signaling in the immune system. Nature Immunology. 2017;**18**:374-384. DOI: 10.1038/ni.3691

[75] Boehi F, Manetsch P, Hottiger MO. Interplay between ADP-ribosyltransferases and essential cell signaling pathways controls cellular responses. Cell Discovery. 2021;7:104

[76] Schindler C, Plumlee C. Inteferons pen the JAK-STAT pathway. Seminars in Cell and Developmental Biology. 2008;**19**:311-318. DOI: 10.1016/j.semcdb.2008.08.010

[77] Villarino AV, Kanno Y, O'Shea JJ. Mechanisms of Jak/STAT signaling in immunity and disease. Journal of Immunology. 2015;**194**:21-27. DOI: 10.4049/jimmunol.1401867

[78] Abboud R, Choi J, Ruminski P, Schroeder MA, Kim S, Abboud CN, et al. Insights into the role of the JAK/STAT signaling pathway in graft-versus-host disease. Therapeutic Advances in Hematology. 2020;**11**:1-13. DOI: 10.1177/2040620720914489

[79] Choi J, Ziga ED, Ritchey J, Collins L, Prior JL, Cooper ML, et al. IFNγR signaling mediates alloreactive T-cell trafficking and GVHD. Blood. 2012;**120**:4093-1043

[80] Lazaryan A, Weisdorf DJ, DeFor T, Brunstein CG, MacMillan ML, Bejanyan N, et al. Risk factors for acute and chronic graft-versus-host disease after allogeneic hematopoietic cell transplantation with umbilical cord blood and matched sibling donors. Biology of Blood and Marrow Transplantation. 2016;**22**:134-140

[81] Schroeder MA, Choi J, Staser K, DiPersio JF. The role of Janus kinase signaling in graft-versus-host disease and graft versus leukemia. Biology of Blood and Marrow Transplantation. 2018;**24**:1125-1134. DOI: 10.1016/j.bbmt.2017.12.797

[82] Spoerl S, Mathew NR, Bscheider M, Schmitt-Graeff A, Chen S, Mueller T, et al. Activity of therapeutic JAK 1/2 blockade in graft-versus-host disease. Blood. 2014;**123**:3832-3842. DOI: 10.1182/blood-2013-12-543736

[83] Betts BC, Abdel-Wahab O, Curran SA, Angelo ETS, Koppikar P, Heller G, et al. Janus kinase-2 inhibition induces durable tolerance to alloantigen by human dendritic cell–stimulated T cells yet preserves immunity to recall antigen. Blood. 2011;**118**:5330-5339. DOI: 10.1182/blood-2011-06-363408

[84] Alexander KA, Flynn R, Lineburg KE, Kuns RD, Teal BE, Olver SD, et al. CSF-1-dependant donor-derived macrophages mediate chronic graft-versus-host disease. Journal of Clinical Investigation. 2014:4266-4280. DOI: 10.1172/JCI75935

[85] Lescoat A, Lelong M, Jeljeli M, Piquet-Pellorce C, Morzadec C, Ballerie A,

et al. Combined anti-fibrotic and anti-inflammatory properties of JAK-inhibitors on macrophages in vitro and in vivo perspectives for scleroderma-associated interstitial lung disease. Biochemical Pharmacology. 2020;**178**:114103. DOI: 10.1016/j.bcp.2020.114103

[86] FDA approves ruxolitinib for acute graft-versus-host disease. 2019 Available from: https://www.fda.gov/drugs/resources-information-approved-drugs/fda-approves-ruxolitinib-acute-graft-versus-host-disease

[87] FDA approves ruxolitinib for chronic graft-versus-host disease. 2021 Available from: https://www.fda.gov/drugs/resources-information-approved-drugs/fda-approves-ruxolitinib-chronic-graft-versus-host-disease

[88] Davis RR, Li B, Yun SY, Chan A, Nareddy P, Gunawan S, et al. Structural Insights into JAK2 Inhibition by Ruxolitinib, Fedratinib, and Derivatives Thereof. The Journal of Medicinal Chemistry. 2021;**64**:2228-2241. DOI: 10.1021/acs.jmedchem.0c01952

[89] You H, Xu D, Zhao J, Li J, Wang Q, Tian X, et al. JAK inhibitors: Prospects in connective tissue diseases. Clinical Reviews in Allergy & Immunology. 2020;**59**:334-351. DOI: 10.1007/s12016-020-08786-6

[90] Schroeder MA, Khoury HJ, Jagasia M, Ali H, Schiller GJ, Staser K, et al. A phase 1 trial of itacitinib, a selective JAK1 inhibitor, in patients with acute graft-versus-host disease. Blood Advances. 2020;**4**:1656-1669. DOI: 10.1182/bloodadvances.2019001043

[91] Dorrington MG, Fraser IDC. NF-κB signaling in macrophages: Dynamics, crosstalk, and signal integration. Frontiers in Immunology. 2019;**10**:705. DOI: 10.3389/fimmu.2019.00705

[92] Oeckinghaus A, Ghosh S. The NF-κB family of transcription factors and its regulation. Cold Spring Harvor Perspectives in Biology. 2009;**1**:a000034. DOI: 10.1101/cshperspect.a00003410.1101/cshperspect.a000034

[93] Siebenlist U, Franzoso G, Brown K. Structure, regulation and function of NF-kappa B. Annual Review of Cell Biology. 1994;**10**:405-455

[94] Ghosh S, Baltimore D. Activation in vitro of NF-kappa B by phosphorylation of its inhibitor I kappa B. Nature. 1990;**344**:678-682.DOI:10.1038/344678a0

[95] Arenzana-Seisdedos F, Turpin P, Rodriguez M, Thomas D, Hay RT, Virelizier JL, et al. Nuclear localization of the IκBα promotes active transport of NF-κB from the nucleus to the cytoplasm. Journal of Cell Science. 1997;**110**:369-378

[96] Karin M, Ben-Neriah Y. Phosphorylation meets ubiquitination: The control of NF-[kappa]B activity. Annual Review of Immunology. 2000;**18**:621-663. DOI: 10.1146/annurev.immunol.18.1.621

[97] Baeuerle PA, Baichwal VR. NF-kappa B as a frequent target for immunosuppressive and anti-inflammatory molecules. Advances in Immunology. 1997;**65**:111-137

[98] MacDonald KPA, Kuns RD, Rowe V, Morris ES, Banovic T, Bofinger H, et al. Effector and regulatory T-cell function is differentially regulated by RelB within antigen-presenting cells during GVHD. Blood. 2007;**109**:5049-5057. DOI: 10.1182/blood-2007-01-067249

[99] Srinivasan M, Flynn R, Price A, Ranger A, Browning JL, Taylor PA, et al. Donor B-cell alloantibody deposition and germinal center formation are required

for the development of murine chronic GVHD and bronchiolitis obliterans. Blood. 2012;**119**:1570-1580. DOI: 10.1182/blood-2011-07-364414

[100] Flynn R, Du J, Veenstra RG, Reichenbach DK, Panoskaltsis-Mortari A, Taylor PA, et al. Increased T follicular helper cells and germinal center B cells are required for cGVHD and bronchiolitis obliterans. Blood. 2014;**123**:3988-3998. DOI: 10.1182/blood-2014-03-562231

[101] Velcade (bortezomib) Information. Available from: https://www.fda.gov/drugs/postmarket-drug-safety-information-patients-and-providers/velcade-bortezomib-information

[102] Koreth J, Kim HT, Lange PB, Poryanda SJ, Reynolds CG, Rai SC, et al. Bortezomib-based immunosuppression after reduced-intensity conditioning hematopoietic stem cell transplantation: Randomized phase II results. Haematologica. 2018;**103**:522-530. DOI: 10.3324/haematol.2017.176859

[103] Pai C-C S, Chen M, Mirsoian A, Grossenbacher SK, Tellez J, Ames E, et al. Treatment of chronic graft-versus-host disease with bortezomib. Blood. 2014;**124**:1677-1688. DOI: 10.1182/blood-2014-02-554279

[104] Martin PJ. Bortezomib for prevention of acute graft- versus-host disease: A conclusion reached. Haematologica. 2018;**103**:377-379. DOI: 10.3324/haematol.2018.188052

[105] Babu D, Fanelli A, Mellone S, Muniswamy R, Wasniewska M, Prodam F, et al. Novel GLI2 mutations identified in patients with Combined Pituitary Hormone Deficiency (CPHD) Evidence for a pathogenic effect by functional characterization. Clinical Endocrinology. 2019;**90**:449-456. DOI: 10.1111/cen.13914

[106] Cochrane CR, Szczepny A, Watkins N, Cain JE. Hedgehog signaling in the maintenance of cancer stem cells. Cancers. 2015;7:1554-1585. DOI: 10.3390/cancers7030851

[107] Chen Y, Struhl G. Dual roles for patched in sequestering and transducing Hedgehog. Cell. 1996;**87**:553-563. DOI: 10.1016/s0092-8674(00)81374-4

[108] Zerr P, Palumbo-Zerr K, Distler A, Tomcik M, Vollath S, Munoz LE, et al. Inhibition of hedgehog signaling for the treatment of murine sclerodermatous chronic graft-versus-host disease. Blood. 2012;**120**:2909-2917. DOI: 10.1182/blood-2012-01-403428

[109] Radojcic V, Pletneva M, Lee CJ, Ivcevic S, Sarantopoulos S, Couriel D. Hedgehog blockade in steroid-refractory sclerotic chronic graft-versus-host disease. British Journal of Haematology. 2021;**195**:e120-e122. DOI: 10.1111/bjh.17657

[110] Du F-Y, Zhou Q-F, Sun W-J, Chen G-L. Targeting cancer stem cells in drug discovery: Current state and future perspectives. World Journal of Stem Cells. 2019;**11**:398-420. DOI: 10.4252/wjsc.v11.i7.398

[111] DeFilipp Z, Nazarian RM, El-Jawahri A, Li S, Brown J, Rio CD, et al. Phase 1 study of the Hedgehog pathway inhibitor sonidegib for steroid-refractory chronic graft-versus-host disease. Blood Advances. 2017;**1**:1919-1922. DOI: 10.1182/bloodadvances.2017011239

[112] Lacroix C, Fish I, Torosyan H, Parathaman P, Irwin JJ, Shoichet BK, et al. Identification of novel smoothened ligands using structure-based docking. PLoS One. 2016;**11**:e0160365. DOI: 10.1371/journal.pone.0160365

[113] Presland RB. Biology of chronic graft-vs-host disease: Immune

mechanisms and progress in biomarker discovery. World Journal of Transplantation. 2020;**6**:608-619. DOI: 10.5500/wjt.v6.i4.608

[114] Mohty M, Apperley JF. Long-term physiological side effects after allogeneic bone marrow transplantation. Hematology/The Education Program of the American Society of Hematology. 2010;**2010**:229-236. DOI: 10.1182/asheducation-2010.1.229

[115] Ozcan U, Yilmaz E, Ozcan L, Furuhashi M, Vaillancourt E, Smith RO, et al. Chemical Chaperones reduce ER stress and restore glucose homeostasis in a mouse model of type 2 diabetes. Science. 2006;**313**:1137-1140. DOI: 10.1126/science.1128294

[116] Kawasaki N, Asada R, Saito A, Kanemoto S, Imaizumi K. Obesity-induced endoplasmic reticulum stress causes chronic inflammation in adipose tissue. Scientific Reports. 2012;**2**:799. DOI: 10.1038/srep00799

[117] Mukai S, Ogawa Y, Urano F, Kudo-Saito C, Kawakami Y, Tsubota K. Novel treatment of chronic graft-versus-host disease in mice using the ER stress reducer 4-phenylbutyric acid. Scientific Reports. 2017;**7**:41939. DOI: 10.1038/srep41939

[118] Kolb PS, Ayaub EA, Zhou W, Yum V, Dickhout JG, Ask K. The therapeutic effects of 4-phenylbutyric acid in maintaining proteostasis. International Journal of Biochemistry & Cell Biology. 2015;**61**:45-52

[119] Inamoto Y, Martin PJ, Paczesny S, Tabellini L, Momin AA, Mumaw CL, et al. Association of plasma CD163 concentration with De Novo-Onset chronic graft-versus-host disease. Biology of Blood and Marrow Transplantation. 2017;**23**:1250-1256. DOI: 10.1016/j.bbmt.2017.04.019

[120] Lim J-Y, Ryu D-B, Lee S-E, Park G, Min C-K. Mesenchymal stem cells (MSCs) attenuate cutaneous sclerodermatous graft-versus-host disease (Scl-GVHD) through inhibition of immune cell infiltration in a mouse model. Journal of Investigative Dermatology. 2017;**137**:1895-1904. DOI: 10.1016/j.jid.2017.02.986

[121] Hong Y-Q, Wan B, Li X-F. Macrophage regulation of graft-vs-host disease. World Journal of Clinical Cases. 2020;**8**:1793-1805

[122] Guerrero-Juarez CF, Dedhia PH, Jin S, Ruiz-Vega R, Ma D, Liu Y, et al. Single-cell analysis reveals fibroblast heterogeneity and myeloid-derived adipocyte progenitors in murine skin wounds. Nature Communications. 2019;**10**:650. DOI: 10.1038/s41467-018-08247-x

[123] LeBleu VS, Neilson EG. Origin and functional heterogeneity of fibroblasts. FASEB Journal. 2020;**34**:3519-3536. DOI: 10.1096/fj.201903188R

[124] Gordon S, Plüddemann A, Estrada FM. Macrophage heterogeneity in tissues: Phenotypic diversity and functions. Immunological Reviews. 2014;**262**:36-55. DOI: 10.1111/imr.12223

[125] Gordon S, Plüddemann A. Tissue macrophages: Heterogeneity and functions. BMC Biology. 2017;**15**:53. DOI: 10.1186/s12915-017-0392-4

[126] Decano JL, Aikawa M. Dynamic macrophages: Understanding mechanisms of activation as guide to therapy for atherosclerotic vascular disease. Frontiers in Cardiovascular Medicine. 2018;**5**:1-12

[127] Mukai S, Ogawa Y, Kawakami Y, Mashima Y, Tsubota K. Inhibition of vascular adhesion protein-1 for treatment

of graft-versus-host disease in mice. FASEB Journal. 2018;**32**:4085-4095. DOI: 10.1096/fj.201700176R

[128] Jalkanen S, Bargatze RF, Toyos J, Butcher EC. Lymphocyte recognition of high endothelium: Antibodies to distinct epitopes of an 85-95-kD glycoprotein antigen differentially inhibit lymphocyte binding to lymph node, mucosal, or synovial endothelial cells. Journal of Cell Biology. 1987;**105**:983-990. DOI: 10.1083/jcb.105.2.983

[129] Foot JS, Yow TT, Schilter H, Buson A, Deodhar M, Findlay AD, et al. PXS-4681A, a potent and selective mechanism-based inhibitor of SSAO/VAP-1 with anti-inflammatory effects in vivo. The Journal of Pharmacology and Experimental Therapeutics. 2013;**347**: 365-374. DOI: 10.1124/jpet.113.207613

[130] Bligt-Lindén E, Pihlavisto M, Szatmári I, Otwinowski Z, Smith DJ, Lázár L, et al. Novel pyridazinone inhibitors for vascular adhesion protein-1 (VAP-1): Old target-new inhibition mode. The Journal of Medicinal Chemistry. 2013;**56**:9837-9848. DOI: 10.1021/jm401372d

[131] Hori K, Sen A, Artavanis-Tsakonas S. Notch signaling at a glance. Journal of Cell Science. 2013;**126**:2135-2140

[132] Tian L, Wu X, Chi C, Han M, Xu T, Zhuang Y. ADAM10 is essential for proteolytic activation of Notch during thymocyte development. International Immunology. 2008;**20**:1181-1187. DOI: 10.1093/intimm/dxn076

[133] Ianni MD, Papa BD, Baldoni S, Tommaso AD, Fabi B, Rosati E, et al. NOTCH and graft-versus-host disease. Frontiers in Immunology. 2018;**9**:1825. DOI: 10.3389/fimmu.2018.01825

[134] Grazioli P, Felli MP, Screpanti I, Campese AF. The mazy case of Notch and immunoregulatory cells. Journal of Leukocyte Biology. 2017;**102**:361-368. DOI: 10.1189/jlb.1VMR1216-505R

[135] Zhang Y, Sandy AR, Wang J, Radojcic V, Shan GT, Tran IT, et al. Notch signaling is a critical regulator of allogeneic CD4$^+$ T-cell responses mediating graft-versus-host disease. Blood. 2011;**117**:299-308. DOI: 10.1182/blood-2010-03-271940

[136] Mochizuki K, Xie F, He S, Tong Q, Liu Y, Mochizuki I, et al. Delta-like ligand 4 identifies a previously uncharacterized population of inflammatory dendritic cells that plays important roles in eliciting allogeneic T cell responses in mice. Journal of Immunology. 2013;**190**:3772-3782. DOI: 10.4049/jimmunol.1202820

[137] Radojcic V, Paz K, Chung J, Du J, Perkey ET, Flynn R, et al. Notch signaling mediated by Delta-like ligands 1 and 4 controls the pathogenesis of chronic GVHD in mice. Blood. 2018;**132**:2188-2200. DOI: 10.1182/blood-2018-03-841155

[138] Poe JC, Jia W, Su H, Anand S, Rose JJ, Tata PV, et al. An aberrant NOTCH2-BCR signaling axis in B cells from patients with chronic GVHD. Blood. 2017;**130**:2131-2145. DOI: 10.1182/blood-2017-05-782466

[139] Tran IT, Sandy AR, Carulli AJ, Ebens C, Chung J, Shan GT, et al. Blockade of individual Notch ligands and receptors controls graft-versus-host disease. Journal of Clinical Investigation. 2013;**123**:1590-1604. DOI: 10.1172/JCI65477

[140] Julian L, Olson M F: Rho-associated coiled-coil containing kinases (ROCK). Small GTPases. 2014;**5**:e29846

[141] Jagasia M, Lazaryan A, Bachier CR, Salhotra A, Weisdorf DJ, Zoghi B, et al.

ROCK2 inhibition with belumosudil (KD025) for the treatment of chronic graft-versus-host disease. Journal of Clinical Oncology. 2021;**39**:1888-1898. DOI: 10.1200/JCO.20.02754

[142] Zanin-Zhorov A, Weiss JM, Nyuydzefe MS, Chen W, Scher JU, Mo R, et al. Selective oral ROCK2 inhibitor down-regulates IL-21 and IL-17 secretion in human T cells via STAT3-dependent mechanism. Proceedings of the National Academy of Sciences of the United States of America. 2014;**111**:16814-16819. DOI: 10.1073/pnas.1414189111

[143] Riches DWH, Backos DS, Redente EF. ROCK and Rho: Promising therapeutic targets to ameliorate pulmonary fibrosis. The American Journal of Pathology. 2015;**185**:909-912. DOI: 10.1016/j.ajpath.2015.01.005

[144] Knipe RS, Tager AM, Liao JK. The Rho kinases: Critical mediators of multiple profibrotic processes and rational targets for new therapies for pulmonary fibrosis. Pharmacological Reviews. 2015;**67**:103-117. DOI: 10.1124/pr.114.009381

[145] Flynn R, Paz K, Du J, Reichenbach DK, Taylor PA, Panoskaltsis-Mortari A, et al. Targeted Rho-associated kinase 2 inhibition suppresses murine and human chronic GVHD through a Stat3-dependent mechanism. Blood. 2016;**127**:2144-2154. DOI: 10.1182/blood-2015-10-678706

[146] FDA approves belumosudil for chronic graft-versus-host disease. doi: https://www.fda.gov/drugs/resources-information-approved-drugs/fda-approves-belumosudil-chronic-graft-versus-host-disease

[147] Sternberg A: Robust Responses With ROCK2 Inhibition Are Seen in Chronic GVHD. 2020. Available from https://www.targetedonc.com/view/robust-responses-with-rock2-inhibition-are-seen-in-chronic-gvhd

[148] Nagumo H, Sasaki Y, Ono Y, Okamoto H, Seto M, Takuwa Y. Rho kinase inhibitor HA-1077 prevents Rho-mediated myosin phosphatase inhibition in smooth muscle cells. The American Journal of Physiology-Cell Physiology. 2000;**278**:C57-C65. DOI: 10.1152/ajpcell.2000.278.1.C57

[149] Naraoka M, Munakata A, Matsuda N, Shimamura N, Ohkuma H. Suppression of the Rho/Rho-kinase pathway and prevention of cerebral vasospasm by combination treatment with statin and fasudil after subarachnoid hemorrhage in rabbit. Translational Stroke Research. 2013;**4**:368-374. DOI: 10.1007/s12975-012-0247-9

[150] Inoue T, Tanihara H. Ripasudil hydrochloride hydrate: Targeting Rho kinase in the treatment of glaucoma. Expert Opinion on Pharmacotherapy. 2017;**18**:1669-1673. DOI: 10.1080/14656566.2017.1378344

[151] Lin C-W, Sherman B, Moore LA, Laethem CL, Lu D-W, Pattabiraman PP, et al. Discovery and preclinical development of netarsudil, a novel ocular hypotensive agent for the treatment of glaucoma. Journal of Ocular Pharmacology and Therapeutics. 2018;**34**:40-51. DOI: 10.1089/jop.2017.0023

[152] Hu Z, Wang C, Sitkoff D, Cheadle NL, Xu S, Muckelbauer JK, et al. Identification of 5H-chromeno[3,4-c]pyridine and 6H-isochromeno[3,4-c]pyridine derivatives as potent and selective dual ROCK inhibitors. Bioorganic & Medicinal Chemistry Letters. 2020;**30**:127474

Chapter 5

Involvement of the Purinergic System in Cell Death in Models of Retinopathies

Douglas Penaforte Cruz, Marinna Garcia Repossi and Lucianne Fragel Madeira

Abstract

Literature data demonstrate already that the presence of adenine nucleotides in the extracellular environment induces cell death that leads to several retinopathies. As said, the objective is to carry out a systematized review of the last decade, relating purinergic signaling to the outcome of cell death and retinopathies. It is possible to identify different mechanisms that occur through the activation of purinergic receptors. The exacerbated activation of the P2X7 receptor is mainly involved in the apoptotic death pathway, and this response is due to the dysregulation of some components in the intracellular environment, such as the Ca^{2+} ion, CD40, MiR-187, and influence of mononuclear macrophages. The A2A receptor is involved in increasing levels of cytokines and promoting inflammatory processes. The data presented can be used as a basis to better understand the mechanisms of death in retinopathies, in addition to proposing therapeutic strategies with the potential to be transposed to several other models.

Keywords: P2x7 receptor, A2a receptor, apoptosis, retina

1. Introduction

The retina is a tissue that is located at the back of the eye and is responsible for converting light stimuli into electrical signals, a process is known as phototransduction, being responsible for the sense of vision [1]. In this tissue, as in many others, cell death is a highly regulated process that is important for maintaining homeostasis, in addition to preserving tissue function by excluding cells whose genome is altered [2]. According to Fricker et al. [3], there are some particularities in cell death in neurons. For example, complexity in nervous system circuitry during development results in programmed cell death of neurons that fail to connect properly. Furthermore, the excitability of neurons causes a high volume of adenosine triphosphate (ATP) and, due to a cytotoxic effect of this molecule in high concentrations, a sensitivity to death in several pathological models [3].

Damage or dysfunction in the retina, through excess cell death, can lead to several pathological conditions that interfere with the normal function of the tissue, compromising the ability to transmit visual stimuli to higher centers. A possible classification of diseases that affect the retina is hereditary retinal dystrophies, a group of pathologies caused by spontaneous genetic alterations that lead to irreversible loss of vision [4]. These diseases have great clinical relevance since they have a high incidence, and in Brazil, the ones that stand out are Non-syndromic Retinitis Pigmentosa, Stargardt's disease, Leber's Congenital Amaurosis, and hereditary retinal syndromic dystrophies [5]. Another known classification is retinopathies, a set of pathologies that affect the retina, which involve damage to the surrounding vasculature [6]. The decrease in the number of blood vessels leads to a process called retinal neovascularization that, occurring in a disorderly way, can deregulate tissue homeostasis or cause the process of retinal detachment [7].

Diseases in which vascular damage occurs are referred to as ischemic retinopathies and include diabetic retinopathy, retinal vein occlusions (RVOs), retinopathy of prematurity (ROP), and sickle-cell retinopathy. Thus, following this definition, other diseases fit this classification as they present stages with vascular damage, such as age-related macular degeneration (AMD), retinal detachment, and glaucoma [8]. Both in the case of retinopathies and retinal dystrophies, most of them remain without a cure or with inadequate treatment and, therefore, new therapeutic strategies are necessary. In this case, the manipulation of molecular pathways can change the course of cell death, and the participation of the purinergic system is of great interest in these cases.

Cell death in non-neural cells has already been described, in which extracellular ATP triggers cell death by binding to P2X7 receptors, which is the main receptor activated in these cases. Upon activation, P2X7 receptors induce large, non-selective membrane pores, which eventually lead to cell death [9–12]. In addition to the evident relationship between the P2X7 receptor and the cell death process, other purinergic receptors participate in this cellular response. The P2X2 and P2X4 receptors, for example, are upregulated in squirrel ischemia models, and the improvement in cell death with the inhibition of these receptors confirms the participation of purinergic signaling [13].

Purinergic signaling mediated by extracellular ATP and adenosine is involved in the induction and protection of cell death in several models of retinal diseases [14]. One of the mechanisms observed in retinal pathologies is the increased expression of purinergic receptors that contribute to high calcium concentrations. P2X receptors act as direct channels for calcium influx and as indirect activators of voltage-gated calcium channels. Meanwhile, activation of P2Y receptors induces a rapid transient release of calcium from internal stores, followed by an influx of calcium. The increased expression, mainly of the P2X7 receptor, is involved in the pathogenesis of several diseases that affect the retina, in which ATP in high concentrations is capable of inducing apoptosis through this purinergic receptor [15].

Studies show that ATP induces apoptosis in embryonic retinal neurons of chicks in culture through the activation of the P2X7 receptor and ionotropic glutamate receptors [16]. Other research also proves that the direct application of ATP to isolated retinas induces the death of cholinergic amacrine cells that express P2X7 receptors [9, 10].

Although ATP is well known for being toxic in high concentrations, and its receptors are involved in several pathologies as inducers of cell death, adenosine receptors are also worth mentioning. When it comes to cell death, the adenosine A2A receptor plays a role [17]. It has already been seen that blocking the A2A receptor controls microglial reactivity [18], delays excitotoxic death of embryonic motor neurons *in vitro* [19], and is able to prevent cell death in ischemic retinas [20]. This evidence makes it clear that this receptor is of great interest when it comes to therapeutic targets to improve the condition of various diseases in the retina and nervous system.

Considering the close connection between the release of adenine nucleotides in the extracellular environment and cell death through the activation of purinergic receptors, it is of great interest to observe the latest studies carried out on the subject. Therefore, the objective of the present review is to present the knowledge obtained from the studies of the last decade, making clear the participation of purinergic signaling in cell death induced by different models of retinopathies.

2. Age-related macular degeneration (AMD)

2.1 Participation of calcium pathways

Age-related macular degeneration (AMD) is the main cause of irreversible vision loss in the elderly in developed countries, being able to affect several cell types in the retina [21, 22]. From the observed damage, numerous studies try to understand the molecular mechanisms involved in this pathology.

The retinal pigment epithelium (RPE) is considered a site of great interest in this pathology, and it is now well known that ATP acts as a key signaling molecule in several cellular processes, including cell death [23]. The P2X7 receptor is also involved in inflammation and oxidative stress in many cell types, and cell death, inflammation, and oxidative stress have been implicated in AMD.

Through the use of apoptotic markers, Yang et al. sought to know whether the presence of ATP, an endogenous P2X receptor agonist, increased the number of cells undergoing apoptosis in human pigment epithelium cell cultures. In cultures treated with ATP, it was possible to observe an increase in the intensity of apoptotic markers when compared to the control. This effect was blocked by the administration of the oxidized P2X7 antagonist (oATP). The selective exogenous P2X7 receptor agonist, 3'-O-(4-benzoyl) benzoyl-ATP (BzATP), was also able to increase apoptosis, but Brillant Blue G (BBG), a P2X7 receptor antagonist, and oATP reverse this effect [24].

Treatment with BAPTA-AM, used to decrease intracellular calcium levels, was able to decrease ATP and BzATP-induced apoptosis, which indicates that Ca^{2+} is an essential component for signaling the P2X7 receptor pathway and continuation of the apoptotic cascade. In summary, the study by Yang et al. provides the first evidence of the presence of functional P2X7 receptors in human pigment epithelium cell cultures and demonstrates that activation of P2X receptors, especially P2X7 receptors, induces Ca^{2+} signaling and apoptosis in these cells [24].

2.2 Participation of oxysterols

Another implication of AMD is the accumulation of drusen, which are extracellular proteolipid deposits, contributing to vision loss in the advanced stages of the disease. These deposits are located between the RPE and Bruch's membrane (inner layer of the choroid) and contain β-amyloid peptide as the main component [25]. Different oxysterols were found in human drusen, which suggests their involvement in AMD. Furthermore, the aggregated form of β-amyloid is well known as an inducer of oxidative stress and cell death [26, 27].

Oliver et al. aimed to highlight the β-amyloid/oxysterols relationship and describe the involvement of the P2X7-pannexin-1 receptor in oxysterol toxicity in human

RPE cell cultures [28]. A link was found between the presence of β-amyloid peptide aggregates and oxysterol levels. Two types of oxysterols, 25-OH and 7-KC seem to play a role in the pathogenesis of AMD through P2X7 activation, but only 25-OH causes pannexin-dependent pore opening in the cell membrane. This pannexin-stimulated pore opening is important in the pathological mechanism of the disease, as it promotes the extravasation of ATP to the extracellular environment, and consequent activation of the P2X7 receptor. Thus, the toxicity of this oxysterol occurs in two ways—increased P2X7 receptor activity and oxidative stress-dependent on pannexin-1, and pannexin-1-independent chromatin condensation [28].

The potential relationship between oxysterols and β-amyloid in AMD supports the notion that oxysterols can be considered as biomarkers of retinal degeneration. Considering the fundamental role of P2X7 receptor activation in oxysterol cytotoxicity, this may be an important target for the development of treatments for this disease [28].

2.3 Participation of alu RNA

Geographic atrophy (GA) is an advanced form of age-related macular degeneration characterized by central loss of vision due to confluent areas of retinal pigment epithelium loss and degeneration of overlying photoreceptors [25]. The DICER1 processing enzyme is specifically reduced in the RPE in eyes with geographic atrophy, as its blockage results in an abundant increase in Alu RNA transcripts (an endogenous retroelement that requires reverse transcriptase for its life cycle), which in turn promotes the cell death of the RPE [29].

More recent studies have identified that the cytotoxicity of Alu RNA in the RPE is mediated by the activation of the NLRP3 inflammasome [30], and it has already been observed that reactive oxygen species (ROS) and the P2X7 receptor are involved in this process in other systems [31, 32]. Therefore, Kerur et al. investigated whether P2X7 signaling was also involved in Alu RNA-induced NLRP3 inflammasome activation, with experiments performed in mouse and human retinal pigment epithelial cell cultures [33].

After transfection of the Alu RNA into the culture media containing the cells of interest, it was seen that NF-kB signaling and P2X7 activation play important roles in Alu RNA-induced inflammasome initiation and activation and RPE degeneration. The authors also suggested, from cell cultures of P2X7 receptor knockout mice, that this receptor is an essential intermediate in the Alu RNA-induced activation of the NLRP3 inflammasome and consequent RPE degeneration. This suggests that manipulating this pathway may be a useful strategy for developing drugs for the treatment of geographic atrophy [33].

In complete agreement with these results, Fowler et al. also investigated the relationship between the P2X7 purinergic receptor and Alu RNA-induced AMD. It was based on what was previously demonstrated, that Alu-derived RNA activates the NLRP3 inflammasome, via the P2X7 receptor, to cause cell death of the retinal pigment epithelium in geographic atrophy [33]. As Alu RNA requires reverse transcriptase for its life cycle, the use of transcriptase inhibitors has been proposed for a definition of other therapeutic alternatives for the disease [34].

After injection of Alu RNA or transfection into human and mouse retinal pigment epithelium cell cultures, Alu RNA was seen to be cytotoxic, as it activates caspase-1 and activates IRAK4 (interleukin-1 receptor-associated kinase 4), whose phosphorylation in these cases leads to degeneration of the pigment epithelium). Alu and LPS, a bacterial compound known to activate inflammatory pathways, activate the NLRP3 inflammasome via activation of the P2X7 receptor. d4T (reverse transcriptase inhibitor—NRTI) acts in a protective manner, preventing caspase-1 activation and IRAK4 phosphorylation. Several approved and clinically relevant NRTIs, including lamivudine (3TC) and abacavir (ABC), prevented the activation of caspase-1, and the Alu RNA-induced inflammasome effect of NLRP3. NRTIs were effective in mouse models of geographic atrophy, choroidal neovascularization, graft-versus-host disease, and sterile liver inflammation [34].

2.4 Abnormal vascular growth

In the neovascular form of age-related macular degeneration, visual loss commonly occurs as a result of the invasion of abnormal blood vessels from the choroidal circulation, that is, choroidal neovascularization (CNV), which induces irreversible damage to the overlying retina. CNV mainly occurs due to dysregulation in the production of endothelial growth factors in the retinal vascular network [35].

Photoreceptor degeneration involves the activation of several regulated cell death signaling pathways that may constitute potential therapeutic targets. ATP has already been discovered as an important extracellular messenger that may contribute to lethal signaling [36]. Thus, Notomi et al. hypothesized that ATP acting via the P2X7 receptor is involved in the pathogenesis of photoreceptor loss in subretinal hemorrhage.

The results suggest that ATP levels in the subretinal space could potentially be higher than those detected in the vitreous because extracellular ATP diffuses into the vitreous cavity from the subretinal space. From the analysis of cell death in cultures of primary retinas in vitro and in a subretinal hemorrhage model in vivo, it was observed that a concentration of 1 mM of ATP triggered an apoptotic process in photoreceptor cells through binding to the P2X7 receptor, while the use of a selective inhibitor of the P2X7 receptor (Brilliant Blue G (BBG)) was able to prevent this effect. These results indicate that extracellular ATP can trigger apoptosis of photoreceptor cells via P2X7 receptor-dependent machinery. Thus, it is shown that pharmacological inhibition of the P2X7 receptor with BBG may result in neuroprotection of photoreceptors in cases of subretinal hemorrhage [37].

The study further suggests that similar severe neurodegenerative pathologies, such as subarachnoid hemorrhage or intracerebral hemorrhage, may be related to elevations in extracellular ATP. In this way, P2X7 receptor antagonists including BBG may have a neuroprotective therapeutic effect in retinal diseases as well as in Central Nervous System diseases with excessive extracellular ATP.

2.5 Infiltration and accumulation of mononuclear phagocytes

Also focusing on damage to photoreceptors in AMD, Hu et al. related this disease to the infiltration and chronic accumulation of mononuclear phagocytes, [38], which in excess lead to neuronal degeneration [39]. It has also been seen that a deficiency in Cx3cr1, a transmembrane chemokine receptor involved in leukocyte adhesion and migration, leads to the accumulation of mononuclear phagocytes, but the mechanism by which this occurs has not yet been well elucidated [38].

From the isolation of bone marrow-derived monocytes that are recruited to the inflammatory site, the expression levels of the P2X7 receptor in these cells were evaluated by flow cytometry. The study confirmed that the accumulation of mononuclear phagocytes in cases of Cx3Cr1 deficiency leads to increased expression of the P2X7 receptor in these cells. The authors observed that, in these situations, P2X7 receptors provoke the opening of pannexin-dependent pores and release ATP to the external environment. This ATP, from the P2X7 receptor, is able to activate inflammasomes which, in turn, are responsible for the maturation and release of interleukin-1 β (IL-1β), responsible for cytotoxicity and increased cell death in photoreceptors. This was confirmed by the ELISA assay, in which IL-1β levels are increased in cases of Cx3Cr1 deficiency [38].

To test whether P2X7 receptor inhibition has a protective effect against death, intravitreal injection of BBG, a selective inhibitor of the P2X7 receptor, was performed. The TUNEL assay showed that the number of apoptotic cells in the photoreceptor layer was reduced after administration of BBG in cases of Cx3Cr1 deficiency. Immunostaining with Iba-1 to quantify inflammation-associated reactive microglia showed that intravitreal injection of BBG was able to protect against inflammation in these cases.

P2X7 receptor inhibitors, therefore, may be a promising therapeutic target to inhibit lesion expansion in cases of Macular Degeneration, as they may prevent RPE cell death, and IL-1β and P2X7 inhibitors may help to prevent RPE cell death. Photoreceptor loss associated with inflammation [38].

3. Diabetic retinopathy

3.1 Damage to the blood-retinal barrier

Diabetic retinopathy is a serious complication of diabetes mellitus. Breach of the blood-retinal barrier (BRB) is a hallmark of diabetic retinopathy, as well as other eye diseases [40]. The human retina contains two BRBs, the inner and the outer, including endothelial cells and retinal pigment epithelial cells, respectively [41]. Maintenance of the physiological structure of retinal cells requires complex cell-to-cell interactions. These interactions occur at special contact sites called cell junctions, which include tight junctions (TJs), adherent junctions (AJs), and gap junctions (GJs) [42].

Knowing this, Platania et al. tested the hypothesis that activation of the P2X7 receptor contributes to the degradation of the inner portion of the BRB, also interfering with the integrity of the endothelial barrier, through the disruption of TJs between endothelial cells of human retinas, in an environment with high concentrations of glucose [43]. Using the bioinformatics program GEO2R, used to identify differentially expressed genes between two groups, P2X7 receptor expression was measured in human retinal endothelial cell cultures. The expression of the P2X7 receptor underwent a significant increase, induced by both the high concentration of glucose and the agonist BzATP, when compared to the control. Furthermore, high glucose induced the activation and release of the pro-inflammatory cytokine IL-1β via P2X7 receptor activation in human retinal endothelial cells. Glucose exposure also caused a decrease in endothelial cell viability and damage to the BRB [43].

It was also seen, by performing the transendothelial electrical resistance assay (TEER) to measure cell membrane integrity and cell-to-cell interactions, that blocking the P2X7 receptor with the drug JNJ47965567 was able to protect retinal endothelial cells against damage induced by high glucose concentrations and protected the blood-retinal barrier. In addition, treatment with JNJ47965567 significantly decreased the expression and release of IL-1β, induced by high glucose. These findings suggest that the P2X7 receptor plays an important role in regulating the integrity of the retinal blood barrier, and blocking this receptor was useful to counteract the damage caused by high glucose concentration in retinal endothelial cells. Thus, the use of P2X7 receptor antagonists may be useful in the treatment of diabetic retinopathy [43].

3.2 Participation of P2X receptors in hyperglycemic retinas

Long-term exposure to high glucose concentration, considered the main factor in the development of diabetic retinopathy, has already been shown to affect extracellular ATP levels in retinal cell cultures [44]. Furthermore, ATP can act as a neurotransmitter in the retina [45, 46], and through activation of plasma membrane receptors, it can increase intracellular calcium concentration. Some of the inflammatory mediators and excitatory neurotransmitters seen in neuronal death in diabetic retinas are released in response to an increase in intracellular Ca^{2+} concentration. Considering this, Pereira et al. sought to investigate whether the exposure of retinal cells from mice grown under high glucose levels could alter the function of P2X receptors [47].

In this study, through the Western Blot assay, it was seen that cultures of rat retinas exposed to high glucose concentration, the following subunits of the P2X receptor were found—P2X2, P2X3, and P2X7, but these did not undergo any significant change in their content when compared to the control. It is noteworthy that in these retinas the P2X4 receptor was affected by the high concentration of glucose, and its expression was reduced [47].

Through the Fura-2 assay (dye used for labeling intracellular calcium) it was shown that intracellular calcium concentrations triggered by the stimulation of P2 receptors are increased in retinal cells of rats cultured at high glucose concentrations, in a model used to simulate the hyperglycemic conditions seen in diabetes. Also using Fura-2, a difference in the pattern of Ca^{2+} concentration based on cell type was noted. In retinal neurons, the increase in intracellular Ca^{2+} concentration was mainly due to the influx of Ca^{2+} through voltage-sensitive calcium channels. In microglial cells, Ca^{2+} influx occurred mainly through P2X receptor channels, although there was also a minor component of increased intracellular Ca^{2+} concentration dependent on calcium release from intracellular stores [47].

These increased calcium responses may be responsible for the increased release of neurotransmitters and/or inflammatory mediators found in diabetic retinas and therefore contribute to retinal neural cell death detected in the early stages of diabetic retinopathy. Since intracellular calcium plays a key role in cell death, inhibition of some purinergic receptors may exert protective effects against retinal neural cell dysfunction or degeneration, and therefore P2 receptors may become a potential therapeutic target for the treatment of early stages of diabetic retinopathy.

3.3 Involvement of the differentiation cluster 40 (CD40)

Capillary degeneration is a hallmark of early diabetic retinopathy. They are the result of loss of retinal endothelial cells and pericytes (perivascular cells essential in maintaining metabolic, mechanical, and signaling functions in microvessels) [48]. Cluster of differentiation 40 (CD40) is required for retinal capillary degeneration in diabetic mice, a process mediated by the death of retinal endothelial cells [49]. However, binding of CD40 on endothelial cells does not normally induce cell death, likely because CD40 activates PI3K/Akt-mediated pro-survival signals [50, 51]. Thus, Portillo et al. aimed to identify a mechanism by which CD40 triggers programmed cell death in human retinal endothelial cell cultures and address this apparent contradiction [52].

Administration of CD40 ligand in primary cultures of human retinal endothelial cells did not significantly alter the percentage of apoptotic cells. Given the close connection between these cells and Müller's glia, we sought to determine whether Müller's glia would indirectly influence the triggering of CD40-mediated cell death. The results showed that CD40 does not exert its effects directly on endothelial cells, but on circulating Müller's glia. It was also seen, by measuring cytokines by the ELISA assay, that CD40 also did not provoke the secretion of IL-β or TNF-α. In fact, CD40-stimulated Müller glia releases ATP into the extracellular medium. By performing a qPCR, it was noted that CD40 also upregulated the expression of the P2X7 receptor on the surface of endothelial cells, making them susceptible to the cell death process mediated by ATP/P2X7.

To obtain *in vivo* results, the authors used a model of induced diabetes in mice. By performing a real-time PCR after CD40 activation, they concluded that these animals upregulated P2X7 in the retina in a CD40-dependent manner when compared to control. Finally, inhibition of the P2X7 receptor (with A-438079) caused a decrease in retinal endothelial cell-cell death [52].

In summary, these studies have uncovered a mechanism by which CD40 enhances cell death of retinal endothelial cells and suggest that CD40 signaling on Müller cells may be an important contributor to vascular injury in diabetic retinopathy. The expression of CD40 was responsible for the secretion of ATP to the extracellular medium, favoring a greater activation of the P2X7 receptor. Increased programmed cell death accompanies these disorders and the P2X7 receptor is consistently seen as pathogenic in these diseases [53]. The findings may be relevant to other diseases caused by CD40, such as atherosclerosis and inflammatory bowel disease. Thus, new therapies can be developed to treat these diseases: blocking CD40 or the P2X7 receptor may prove to be effective alternatives in the treatment of diabetic retinopathy [52].

Growing evidence indicates that chronic inflammation is important for the development of diabetic retinopathy [54, 55]. TNF-α and IL-1β are pro-inflammatory molecules upregulated in this disease [56]. In addition to macrophages/microglia, Müller's glia (the main retinal microglia) become dysfunctional in diabetes and contribute to the development of diabetic retinopathy. Since CD40 deficiency impairs this process and prevents diabetic retinopathy [52, 57], Portillo et al. sought to elucidate the mechanisms by which this response occurs [58].

The study carried out showed, through real-time PCR, an increase in CD40 expression by Müller's glia in a mouse model of diabetic retinopathy. Additionally, it was seen that CD40 binding on Müller's glia triggered phospholipase C-dependent release of ATP. This release provoked activation of P2X7 receptors, resulting in the release of TNF-α and IL-1β by macrophages. To better prove the role of the P2X7 receptor in this process, mice that do not express the P2X7 receptor and mice treated with a P2X7 inhibitor were protected from the increase in the concentration of TNF-α, IL-1β, ICAM-1, and NOS2 induced by diabetes, thus preventing the inflammatory process and cell death [58].

The observed effects are relevant in vivo because TNF-α is up-regulated in microglia/macrophages of diabetic mice that express CD40 on Müller glia and mice treated with BBG (P2X7 receptor inhibitor) are protected from diabetes-induced upregulation of TNF-α and IL-1β. This protection prevents the inflammatory process that normally accompanies the release of these cytokines, alleviating the pathological effects of diabetic retinopathy. Thus, this study indicates that CD40 in Müller's glia is sufficient to up-regulate retinal inflammatory markers. Furthermore, CD40 appears to promote experimental diabetic retinopathy and Müller's glia orchestrates inflammatory responses in myeloid cells via a CD40-ATP-P2X7 pathway [58].

3.4 Therapeutic potential of the A2A receptor

Diabetic retinopathy is one of the main complications of diabetes mellitus and one of the main causes of blindness. The pathogenesis of diabetic retinopathy is accompanied by chronic low-grade inflammation. Adenosine is a neuromodulator of the central nervous system that exerts its actions through the activation of its four receptors: A1, A2A, A2B, and A3. Some reports demonstrate that the microglial cell response can be altered by A2A receptor antagonists in the different brain and retinal diseases [59, 60]. Therefore, Aires et al. sought to find out whether blocking the A2A receptor can provide protection to the retina by modulating microglial reactivity [61].

Through the use of specific immunomarkers, it was observed that the number of reactive microglia was increased in the retina of mice in a model of induced diabetes. Intravitreal injection of SCH 58261, the A2A receptor antagonist significantly decreased microglial reactivity in the retinas of diabetic animals. The ELISA assay confirmed that, accompanied by this decrease in microglial activity, treatment with the A2A receptor antagonist was able to decrease the levels of TNF and IL-1β cytokines, also demonstrating the ability to control inflammatory processes [61].

In addition to these results, the injection of SCH 58261 was able to decrease the levels of reactive oxygen species. The TUNEL assay confirmed the neuroprotective potential of this inhibitor, demonstrated in the fall of apoptotic cells in the retina of mice in vivo, and the prevention of cell death preserved the thickness of the retinal tissue. Finally, regarding the vascular damage characteristic of Diabetic Retinopathy, it was seen that the inhibition of the A2A receptor contributes to the preservation of the integrity of the blood-retinal barrier. All these data demonstrate the therapeutic potential of A2A receptor antagonists for the treatment of diabetic retinopathy [61].

4. Photoreceptors' degeneration

Photoreceptor degeneration involves the activation of several cell death pathways that may constitute potential therapeutic targets, and an alternative for the inhibition of death pathways is to intercept death, such as the activation of caspases. Among the seven mammalian P2X receptors, the P2X7 receptor has the highest affinity for ATP [36]. Thus, extracellular ATP can induce apoptotic and/or necrotic cell death, acting on the P2X7 receptor [62]. Taking into account the therapeutic possibility of P2X7 receptor inhibitors (such as Brillant Blue G) [11, 63], Notomi et al. decided to investigate the pathogenic implications of the P2X7 receptor in the pathological loss of photoreceptors in mice, as well as the therapeutic utility of BBG in this context [64].

By administering ATP or BzATP (selective P2X7 receptor agonist) in primary cultures of mouse retinal cells, it was seen that stimulation of the P2X7 receptor with these ligands could directly mediate the cell death pathway in photoreceptors. Inhibition of the P2X7 receptor with BBG or KN-62 was able to prevent photoreceptor cell death, confirming the role of this purinergic receptor in this process. The pathway followed after activation of the P2X7 receptor was also demonstrated, in which photoreceptor cell death occurred through calcium influx (observed through the use of the calcium marker Fluo-4 AM) and caspase-8 activation, suggesting a potential mechanism to an extrinsic pathway mediated by the P2X7 receptor [64].

Intravitreal injections of BzATP administered to mice showed that this specific P2X7 receptor agonist has the potential to induce retinal cell death in vivo. Inhibition of the P2X7 receptor proved to be effective in preventing cell death and preserving photoreceptors. Furthermore, blocking the P2X7 receptor indirectly inhibits the caspases of the mitochondrial cell death pathway in retinal cell cultures. Together, these results clarify some of the mechanisms of cell death induced by the binding of ATP to the P2X7 receptor and how antagonists, especially BBG, have clear relevant therapeutic effects that can be transferred to other models of neurodegenerative diseases, having neuroprotective potential that are also relevant. Applies to photoreceptors [64].

5. Glaucoma

5.1 Ischemia-induced damage

As a chronic neurodegenerative condition, glaucoma is characterized by the loss of retinal ganglion cells, resulting in progressive optic neuropathy and consequent visual field loss. Reduced blood flow to the optic nerve region and consequent ischemia has been suggested as a mechanism of ganglion cell death in glaucoma [65, 66].

There is currently considerable interest in the P2X7 receptor in mediating neurodegeneration, with increasing evidence indicating its role in chronic disease [67, 68]. Some studies have also provided evidence that the P2X7 receptor may play a role in glaucoma-induced death [69–71]. Taking this into account, Niyadurupola et al. sought to determine whether stimulation of ischemia-induced death in the ganglion cell layer is mediated by P2X7 in the human retina [12].

As a result, it was seen that stimulation of the P2X7 receptor by the selective agonist BzATP induced cell death in ganglion layer cells in organotypic cultures of human retinas, which was inhibited by the P2X7 receptor inhibitor (BBG). In addition, the ischemia caused to cells in culture led to the loss of retinal ganglion cells, and this effect was also inhibited by BBG, which suggests the participation of the P2X7 receptor in the observed degeneration. Finally, it was possible to locate the P2X7 receptor in the outer and inner plexiform layers of the retina, and the ganglion cells also expressed the mRNA encoding the P2X7 receptor protein [12].

All these data confirm the great importance of this purinergic receptor in the retina and its relationship with glaucoma, since the stimulation of the P2X7 receptor can mediate the death of retinal ganglion cells and that this mechanism plays a role in ischemia-induced neurodegeneration in the human retina. In addition, the therapeutic potential of P2X7 receptor inhibitors is clear, with the aim of preventing cell death [12].

5.2 NMDA-induced damage

It is known that glutamate receptor stimulation by excess glutamate during hypoxia [72] and ischemia-reperfusion [73] is toxic to neurons. Activation of the N-methyl-D-aspartic acid (NMDA) receptor, a subtype of glutamate receptor, is followed by a large influx of Ca^{2+}. This excess of intracellular Ca^{2+} is predominantly involved in neuronal excitotoxicity processes and is considered one of the mechanisms of glaucoma-induced neuronal cell death [71]. Furthermore, some studies have also suggested that the P2X7 receptor plays a role in retinal ganglion cell death induced by high ocular pressure [69, 74, 75]. So, Sakamoto et al. sought to examine the role of P2X7 receptors in NMDA-induced retinal damage in mice in vivo [76].

The results obtained in the study demonstrate that, as expected, the intravitreal injection of the P2X7 receptor agonist (BzATP) induces cell death in the rat retina, an effect that was prevented by the administration of receptor antagonists (A438079 and Brillant Blue G). After this confirmation, it was evaluated whether the NMDA receptor produces its toxic effects through this receptor. Injections of the P2X receptor inhibitors, A438079 and BBG, were able to reduce the deleterious effects of NMDA, decreasing the number of apoptotic cells in cases of glaucoma induced through the NMDA receptor, confirming the neuroprotective effect of these drugs on the retina against toxicity of the drug. Glutamate. Finally, the immunohistochemistry technique was performed to determine the distribution pattern of the P2X7 receptor in mouse retinas. The results indicated that P2X7 receptors were expressed in the somatic region of RGCs that had DAPI-labeled nuclei in the ganglion cell layer and in the inner and outer plexiform layers [76].

These results then showed for the first time that P2X7 receptor activation is, at least in part, involved in NMDA-induced retinal damage. In summary, the study authors demonstrate the possibility that P2X7 receptor antagonists are effective in preventing retinal diseases caused by glutamate excitotoxicity [76].

5.3 Participation of miR-187

MicroRNAs (miRNAs) are a class of non-coding RNAs that regulate transcription and translation of target genes by interacting with the 3′-untranslated region of the target gene (3'-UTR), thus mediating the pathogenesis of multiple human diseases [77]. Previous studies confirmed that miR-187 promoted retinal ganglion cell survival and decreased apoptosis of these cells in human ganglion cell culture incubated with TGF-β, suggesting a protective role of miR-187 against glaucoma [78]. Given the role of miR-187 and the P2X7 receptor in glaucoma, Zhang et al. sought to know whether there is a functional correlation between miR-187 and the P2X7 receptor in apoptosis in a mouse retinal ganglion cell culture-induced model of glaucoma [79].

The results showed that high pressure-induced oxidative stress in retinal ganglion cells was accompanied by a decrease in miR-187 expression and an increase in P2X7 receptor expression. It was also found that miR-187 down-regulated P2X7 receptor expression in ganglion cells, and this inhibition was able to inhibit oxidative stress and apoptosis in these cells [80]. These data demonstrated that miR-187/P2X7 signaling is involved in retinal cell apoptosis, at least in part, through oxidative stress activation. In vitro experiments showed that both miR-187 and the P2X7 receptor were upregulated in the retina of mouse models of chronic ocular hypertension [79]. Thus, miR-187 and the P2X7 receptor promise to be a new target for the prevention and treatment of ophthalmic neurodegenerative diseases.

6. Retinopathy of prematurity (ROP)

Retinopathy of prematurity (ROP) is a disease that can cause blindness in very low birth weight babies and remains a leading cause of childhood blindness in many parts of the world [81, 82]. As a disease that mainly affects the retinal vasculature, existing treatments for this disease, such as anti-VEGF therapy, can have adverse effects, compromising the development of blood vessels and leading to peripheral blindness [80]. The therapeutic potential provided by the antagonism of the purinergic A2A receptor has already been verified, and it may represent a new and promising pharmacological strategy to control pathological retinal angiogenesis under ROP conditions. This strategy avoids the onset of negative effects observed in the anti-VEGF strategy, as it alters molecular mechanisms without compromising the maintenance of the vasculature or the formation of new blood vessels [83]. That said, Zhou et al. sought to extend this potential of A2A receptor inhibition, using it as a therapeutic strategy to selectively control pathological retinal neovascularization, in a model of oxygen-induced retinopathy leading to ROP [84].

To verify whether the use of the A2A receptor inhibitor would alter the retinal vasculature under physiological conditions, KW6002 (A2A receptor inhibitor) was administered intraperitoneally in mice of the C57BL/6 strain. Repeated exposure to KW6002 did not alter the mice's normal retinal vasculature, showing that the treatment has no unwanted effects. After that, immunohistochemistry and immunofluorescence assays showed that, in a model of induced RDP in mice, the administration of KW6002 reduced avascular areas and neovascularization, with apoptosis and cell proliferation also reduced, and astrocyte functions increased. Thus, KW6002 treatment increased astrocyte participation and reduced cell proliferation and apoptosis to confer protection against pathological angiogenesis in ROP [84].

7. Retina detachment

Retinal detachment involves the separation of the sensorineural retina (responsible for receiving and conducting light stimuli to the higher centers of vision) from the retinal pigment epithelium. Direct contact between the retina and the pigment epithelium is essential for its normal function, and detachment can lead to irreversible vision loss [85]. A2A adenosine receptor signaling has been shown to be neuroprotective in some models of retinal damage, but its role in neuronal survival during retinal detachment is unclear. Therefore, Gao et al. sought to modulate the A2A receptor-dependent signaling cascade and observe whether there would be changes in the rate of photoreceptor apoptosis [86].

In a mouse model of retinal detachment, A2A receptor expression was determined from real-time PCR and Western blot assays. It was found that A2A receptor protein was detected in the ganglion cell layer, the inner and outer plexiform layers, and the inner nuclear layer after the retinal detachment protocol. It is worth noting that this receptor was predominantly expressed in microglia and in Müller's glia [86].

The role of the A2A receptor in different models of pathologies is quite controversial, and the cellular response followed may favor cell death or neuroprotection. However, the effect caused by this receptor in cases of retinal detachment had not been well elucidated so far. Thus, in this study, intravitreal injection of the drug ZM241385, a selective antagonist of the A2A receptor, was performed. Through immunofluorescence assays using specific markers, it was seen that blockade of the A2A receptor inhibited microglia reactivity after the triggering of retinal detachment, accompanied by a reduction of microglial proliferation. The drug also decreased the expression of GFAP (reactive gliosis marker) and decreased the expression of the inflammatory cytokine IL-1β. Furthermore, by performing a specific measurement assay for oxidative stress, it was seen that inhibition of the A2A receptor reduced the amount and production of reactive oxygen species in detached retinas [86].

Finally, through the TUNEL assay, the rate of apoptotic cells in the retina after the induction of retinal detachment was evaluated. The administration of ZM241385 was able to prevent the loss of photoreceptors caused by the high concentration of reactive oxygen species triggered after retinal detachment, further reinforcing the role of A2A receptor inhibition in the control of neuroinflammation. Thus, the involvement of the purinergic A2A receptor in the pathogenesis of retinal detachment was confirmed, making it a promising therapeutic target in the treatment of the pathology [86].

8. Transitory retina ischemia

Transient retinal ischemia refers to a pathological condition that involves loss of blood supply to the retina, resulting in cell damage and death from lack of oxygen supply [87]. It has already been seen that a mechanism for triggering this pathological condition is microglia-mediated neuroinflammation, raising the hypothesis that the control of microglial reactivity may provide neuroprotection. Furthermore, it has already been seen that inhibition of the A2A receptor led to neuroprotection from microglial control in cases of ischemia [20]. Taking this into account, Boia et al. aimed to investigate the therapeutic potential of oral administration of the A2A receptor antagonist and the effects of caffeine ingestion (adenosine receptor antagonist) against neuroinflammation and cell death in a model of ischemia caused by intraocular pressure. in mice [60].

Knockout mice for the A2A receptor were used to evaluate the effects of the absence of this receptor in relation to the control in cases of ischemia. From the quantification of TNF and IL-1β levels by the ELISA assay, it was seen that IL-1β levels were not altered, but TNF levels were significantly reduced in A2A knockout retinas when compared to Wild-Type animals. Likewise, the use of the A2A receptor antagonist (KW6002) caused a reduction in TNF levels. In addition, labeling of reactive microglial activity was also found to be decreased when the A2A receptor was inhibited by the drug KW6002. The TUNEL assay confirmed the neuroprotection caused by the use of the A2A receptor inhibitor since the number of apoptotic cells was found to be reduced in relation to the control group [60].

Focusing on caffeine, which is an adenosine receptor antagonist, we quantified the expression of the same cytokines seen previously through qPCR and ELISA assay. In ischemic retinas, the acute administration of the substance was not able to change the levels of TNF or IL-1β; however, the transcriptional levels of the two cytokines were found to be elevated after 24 hours of administration. Regarding microglial activity, caffeine showed dichotomous results: after 24 hours, caffeine increased microglial reactivity, inflammatory response, and ischemia-induced cell death compared to the control group. However, at 7 days of reperfusion, caffeine administration decreased microglia reactivity and reduced levels of pro-inflammatory cytokines and cell death. This indicates that prolonged treatment with caffeine induces the beneficial effects presented [60].

9. Conclusion

Purinergic signaling has already been shown to be important for several cellular processes in different organs and systems. The present study showed the relevance of purinergic receptors in signaling cell death pathways in the retina, and the cytotoxic effects can be applied in the various retinopathies addressed (**Figure 1**). When it comes to cell death, the participation of the P2X7 receptor in cytotoxic and inflammatory processes is clear. Thus, despite the search to encompass the entire purinergic system, P2X7 and A2A receptors were the most found when it comes to cell death in the retina.

In the studies analyzed in this review, it is possible to recognize different cellular mechanisms that occur through the activation of the P2X7 receptor. Dependence on the influx of Ca^{2+} ions after receptor activation was present, since the lack of these ions prevents the apoptotic cascade from occurring, and this pattern was present in models of Macular Degeneration and Diabetic Retinopathy. In AMD models, other mechanisms by which the activation of the P2X7 receptor can act can also be observed, such as the increase in the levels of oxysterols, Alu RNA and the infiltration of mononuclear macrophages. All these factors contribute to the onset of cytotoxic effects and the initiation of cell death.

Cell death induced by purinergic signaling also extends to other pathologies: in Diabetic Retinopathy, it is noted that, in addition to calcium-dependent signaling, the P2X7 receptor is also involved in damage to the blood-retinal barrier (BRB), and damage mediated by the differentiation cluster 40 (CD40); in cases of glaucoma, the ischemic injury induced by NMDA or involving the MicroRNA MiR-187 has its toxic effects also mediated by this same receptor.

We can also see that the adenosine A2A receptor plays an important role in triggering several diseases, mainly in alterations of vascular integrity. This receptor has been

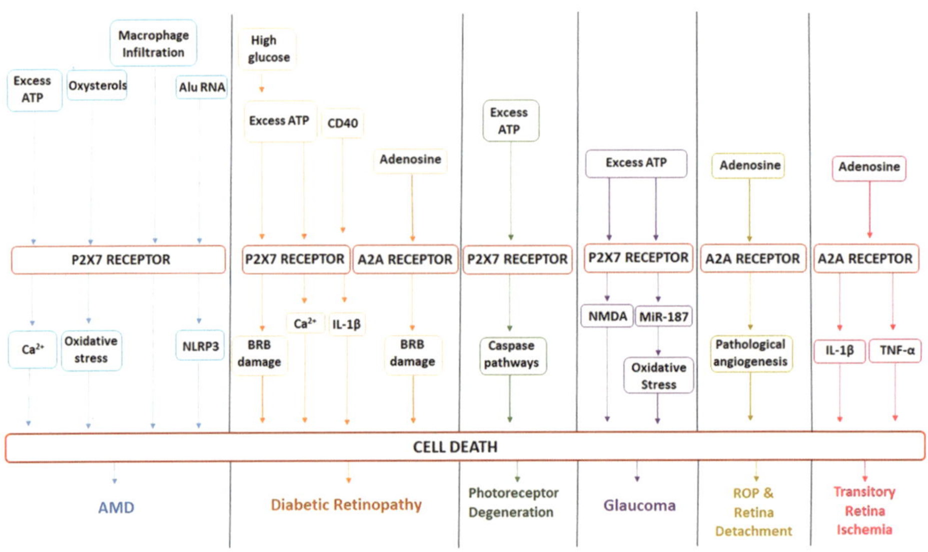

Figure 1.
Synthesis of the involvement of the P2X7 receptor in the presented retinopathies. In the studies analyzed, it was possible to recognize different cellular mechanisms that occur through activation of the P2X7 and A2A receptors, each of them leading to the final outcome, which is cell death.

seen to be involved in increasing cytokine levels in pathological conditions such as TGF and IL-1B; also in triggering microglial reactivity and promoting inflammatory processes.

With the results obtained from this analysis, it is clear that more and more studies claim that the exacerbated activation of the P2X7 receptor by extracellular ATP is highly involved mainly in the apoptotic pathway of cell death. When it comes to possible therapeutic targets for the diseases addressed, it is interesting to consider P2X7 receptor inhibitors for this purpose, as some of the studies show the effectiveness of these inhibitors in improving the effects observed in each model. Thus, the data presented here can be taken as a basis to better understand the mechanisms of death in various retinopathies, in addition to proposing therapeutic strategies with the potential to be transposed to several other models.

Author details

Douglas Penaforte Cruz, Marinna Garcia Repossi and Lucianne Fragel Madeira*
Department of Neurobiology at the Fluminense Federal University, Institute of Biology, Niterói, Rio de Janeiro, Brazil

*Address all correspondence to: lfragel@id.uff.br

References

[1] Vinberg F, Chen J, Kefalov VJ. Regulation of calcium homeostasis in the outer segments of rod and cone photoreceptors. Progress in Retinal and Eye Research. 2018;**67**:87-101. DOI: 10.1016/j.preteyeres.2018.06.001

[2] D'Arcy MS. Cell death. A review of the major forms of apoptosis, necrosis and autophagy. Cell Biology International. 2019;**43**(6):582-592. DOI: 10.1002/cbin.11137

[3] Fricker M et al. Neuronal cell death. Physiological Reviews. 2018;**98**(2):813-880. DOI: 10.1152/physrev.00011.2017

[4] Ziccardi L et al. Molecular sciences gene therapy in retinal dystrophies. International Journal of Molecular Sciences. 2019;**20**(22):5722

[5] Motta FL, Martin RP, Filippelli-Silva R, Salles MV, Sallum JMF. Relative frequency of inherited retinal dystrophies in Brazil. Scientific Reports. 2018;**8**(1):1-9. DOI: 10.1038/s41598-018-34380-0

[6] Torpy JM. Retinopathy. The Journal of the American Medical Association (JAMA). 2007;**298**(8):944. DOI: 10.1001/jama.298.8.944

[7] Campochiaro PA. Molecular pathogenesis of retinal and choroidal vascular diseases. Progress in Retinal and Eye Research. 2015;**49**:67-81. DOI: 10.1016/j.preteyeres.2015.06.002

[8] Ip M, Hendrick A. Retinal vein occlusion review. Asia-Pacific Journal of Ophthalmology. 2018;**7**(1):40-45. DOI: 10.22608/APO.2017442

[9] Resta V, Novelli E, Di Virgilio F, Galli-Resta L. Neuronal death induced by endogenous extracellular ATP in retinal cholinergic neuron density control. Development. 2005;**132**(12):2873-2882. DOI: 10.1242/dev.01855

[10] Zhang X, Zhang M, Laties AM, Mitchell CH. Stimulation of P2X7 receptors elevates Ca^{2+} and kills retinal ganglion cells. Investigative Ophthalmology and Visual Science. 2005;**46**(6):2183-2191. DOI: 10.1167/iovs.05-0052

[11] Peng W et al. Systemic administration of an antagonist of the ATP-sensitive receptor P2X7 improves recovery after spinal cord injury. Proceedings of the National Academy of Sciences of the United States of America. 2009;**106**(30):12489-12493. DOI: 10.1073/pnas.0902531106

[12] Niyadurupola N, Sidaway P, Ma N, Rhodes JD, Broadway DC, Sanderson J. P2X7 receptor activation mediates retinal ganglion cell death in a human retina model of ischemic neurodegeneration. Investigative Ophthalmology & Visual Science. 2013;**54**(3):2163-2170

[13] Cavaliere F et al. Up-regulation of P2X2, P2X4 receptor and ischemic cell death: Prevention by P2 antagonists. Neuroscience. 2003;**120**(1):85-98. DOI: 10.1016/S0306-4522(03)00228-8

[14] Khakh BS, Alan North R. P2X receptors as cell-surface ATP sensors in health and disease. Nature. 2006;**442**(7102):527-532. DOI: 10.1038/nature04886

[15] Reichenbach A, Bringmann A. Purinergic signaling in retinal degeneration and regeneration. Neuropharmacology. 2016;**104**:194-211

[16] Anccasi RM, Ornelas IM, Cossenza M, Persechini PM, Ventura ALM. ATP induces the death of developing avian retinal neurons in culture via activation of P2X7 and glutamate receptors. Purinergic Signal. 2013;**9**(1):15-29. DOI: 10.1007/s11302-012-9324-5

[17] Seven YB, Simon AK, Sajjadi E, Zwick A, Satriotomo I, Mitchell GS. Adenosine 2A receptor inhibition protects phrenic motor neurons from cell death induced by protein synthesis inhibition. Physiology & Behavior. 2017;**176**(12):139-148. DOI: 10.1016/j.expneurol.2019.113067.Adenosine

[18] Orr AG, Orr AL, Li XJ, Gross RE, Traynelis SF. Adenosine A2A receptor mediates microglial process retraction. Nature Neuroscience. 2009;**12**(7):872-878. DOI: 10.1038/nn.2341

[19] Veréb Z et al. Functional and molecular characterization of ex vivo cultured epiretinal membrane cells from human proliferative diabetic retinopathy. BioMed Research International. 2013;**2013**:1-14. DOI: 10.1155/2013/492376

[20] Madeira MH et al. Selective A2A receptor antagonist prevents microglia-mediated neuroinflammation and protects retinal ganglion cells from high intraocular pressure-induced transient ischemic injury. Translational Research. 2016;**169**:112-128. DOI: 10.1016/j.trsl.2015.11.005

[21] Ferris FL, Fine SL, Hyman L. Age-related macular degeneration and blindness due to neovascular maculopathy. Archives of Ophthalmology. 1984;**102**(11):1640-1642. DOI: 10.1001/archopht.1984.01040031330019

[22] Mitchell P, Liew G, Gopinath B, Wong TY. Age-related macular degeneration. Lancet. 2018;**392**(10153):1147-1159. DOI: 10.1016/S0140-6736(18)31550-2

[23] Schwiebert EM, Zsembery A. Extracellular ATP as a signaling molecule for epithelial cells. Biochimica et Biophysica Acta—Biomembranes. 2003;**1615**(1-2):7-32. DOI: 10.1016/S0005-2736(03)00210-4

[24] Yang D, Elner SG, Clark AJ, Hughes BA, Petty HR, Elner VM. Activation of P2X receptors induces apoptosis in human retinal pigment epithelium. Investigative Ophthalmology & Visual Science. 2011;**52**(3):1522-1530

[25] Buschini E, Piras A, Nuzzi R, Vercelli A. Age related macular degeneration and drusen: Neuroinflammation in the retina. Progress in Neurobiology. 2011;**95**(1):14-25. DOI: 10.1016/j.pneurobio.2011.05.011

[26] Gschwind M, Huber G. Apoptotic cell death induced by β-Amyloid1-42 peptide is cell type dependent. Journal of Neurochemistry. 1995;**65**(1):292-300. DOI: 10.1046/j.1471-4159.1995.65010292.x

[27] Bruban J et al. Amyloid-β(1-42) alters structure and function of retinal pigmented epithelial cells. Aging Cell. 2009;**8**(2):162-177. DOI: 10.1111/j.1474-9726.2009.00456.x

[28] Olivier E et al. P2X7-pannexin-1 and amyloid β-induced oxysterol input in human retinal cell: Role in age-related macular degeneration? Biochimie. 2016;**127**:70-78. DOI: 10.1016/j.biochi.2016.04.014

[29] Kaneko H et al. DICER1 deficit induces Alu RNA toxicity in age-related macular degeneration. Nature. 2011;**471**(7338):325-332. DOI: 10.1038/nature09830

[30] Tarallo V et al. DICER1 loss and Alu RNA induce age-related macular degeneration via the NLRP3 inflammasome and MyD88. Cell. 2012;**149**(4):847-859. DOI: 10.1016/j.cell.2012.03.036

[31] Bartlett R, Yerbury JJ, Sluyter R. P2X7 receptor activation induces reactive oxygen species formation and cell death in murine eoc13 microglia. Mediators of Inflammation. 2013;**2013**. DOI: 10.1155/2013/271813

[32] Cruz CM, Rinna A, Forman HJ, Ventura ALM, Persechini PM, Ojcius DM. ATP activates a reactive oxygen species-dependent oxidative stress response and secretion of proinflammatory cytokines in macrophages. The Journal of Biological Chemistry. 2007;**282**(5):2871-2879. DOI: 10.1074/jbc.M608083200

[33] Kerur N et al. TLR-independent and P2X7-dependent signaling mediate Alu RNA-induced NLRP3 inflammasome activation in geographic atrophy. Investigative Ophthalmology and Visual Science. 2013;**54**(12):7395-7401. DOI: 10.1167/iovs.13-12500

[34] Fowler BJ et al. Nucleoside reverse transcriptase inhibitors possess intrinsic anti-inflammatory activity. Science. 2014;**346**(6212):1000-1003. DOI: 10.1126/science.1261754

[35] Grossniklaus HE, Green WR. Choroidal neovascularization. American Journal of Ophthalmology. 2004;**137**(3):496-503. DOI: 10.1016/j.ajo.2003.09.042

[36] Dubyak GR, El-Moatassim C. Signal transduction via P2-purinergic receptors for extracellular ATP and other nucleotides. The American Journal of Physiology. 1993;**265**(3):1. DOI: 10.1152/ajpcell.1993.265.3.c577

[37] Notomi S et al. Dynamic increase in extracellular ATP accelerates photoreceptor cell apoptosis via ligation of P2RX7 in subretinal hemorrhage. PLoS One. 2013;**8**(1):e53338

[38] Hu SJ et al. Upregulation of P2RX7 in Cx3cr1-deficient mononuclear phagocytes leads to increased interleukin-1β secretion and photoreceptor neurodegeneration. The Journal of Neuroscience. 2015;**35**(18):6987-6996. DOI: 10.1523/JNEUROSCI.3955-14.2015

[39] Cardona AE et al. Control of microglial neurotoxicity by the fractalkine receptor. Nature Neuroscience. 2006;**9**(7):917-924. DOI: 10.1038/nn1715

[40] Klaassen I, Van Noorden CJF, Schlingemann RO. Molecular basis of the inner blood-retinal barrier and its breakdown in diabetic macular edema and other pathological conditions. Progress in Retinal and Eye Research. 2013;**34**(Feb):19-48. DOI: 10.1016/j.preteyeres.2013.02.001

[41] Díaz-Coránguez M, Ramos C, Antonetti DA. The inner blood-retinal barrier: Cellular basis and development. Vision Research. 2017;**139**:123-137. DOI: 10.1016/j.visres.2017.05.009

[42] Farquhar MG. Junctional complexes in various epithelia. The Journal of Cell Biology. 1963;**17**(2):375-412. DOI: 10.1083/jcb.17.2.375

[43] Platania CBM et al. Blood-retinal barrier protection against high glucose damage: The role of P2X7 receptor. Biochemical Pharmacology. 2019;**168**:249-258

[44] Costa G, Pereira T, Neto AM, Cristóvão AJ, Ambrósio AF, Santos PF. High glucose changes extracellular adenosine triphosphate levels in rat

retinal cultures. Journal of Neuroscience Research. 2009;**87**(6):1375-1380. DOI: 10.1002/jnr.21956

[45] Santos PF, Caramelo OL, Carvalho AP, Duarte CB. Characterization of ATP release from cultures enriched in cholinergic amacrine-like neurons. Journal of Neurobiology. 1999;**41**(3):340-348. DOI: 10.1002/(SICI)1097-4695(19991115)41:3<340::AID-NEU3>3.0.CO;2-8

[46] Ward MM, Puthussery T, Fletcher EL. Localization and possible function of P2Y4 receptors in the rodent retina. Neuroscience. 2008;**155**(4):1262-1274. DOI: 10.1016/j.neuroscience.2008.06.035

[47] Pereira TDOS, da Costa GNF, Santiago ARS, Ambrósio AF, dos Santos PFM. High glucose enhances intracellular Ca^{2+} responses triggered by purinergic stimulation in retinal neurons and microglia. Brain Research. 2010;**1316**:129-138. DOI: 10.1016/j.brainres.2009.12.034

[48] Mizutani M, Kern TS, Lorenzi M. Accelerated death of retinal microvascular cells in human and experimental diabetic retinopathy. The Journal of Clinical Investigation. 1996;**97**(12):2883-2890. DOI: 10.1172/JCI118746

[49] Portillo JAC et al. CD40 promotes the development of early diabetic retinopathy in mice. Diabetologia. 2014;**57**(10):2222-2231. DOI: 10.1007/s00125-014-3321-x

[50] Deregibus MC, Buttiglieri S, Russo S, Bussolati B, Camussi G. CD40-dependent activation of phosphatidylinositol 3-kinase/Akt pathway mediates endothelial cell survival and in vitro angiogenesis. The Journal of Biological Chemistry. 2003;**278**(20):18008-18014. DOI: 10.1074/jbc.M300711200

[51] Scherl A et al. Functional proteomic analysis of human nucleolus. Molecular Biology of the Cell. 2002;**13**(November):4100-4109. DOI: 10.1091/mbc.E02

[52] Portillo JAC, Corcino YL, Dubyak GR, Kern TS, Matsuyama S, Subauste CS. Ligation of CD40 in human müller cells induces P2x7receptor–dependent death of retinal endothelial cells. Investigative Ophthalmology and Visual Science. 2016;**57**(14):6278-6286. DOI: 10.1167/iovs.16-20301

[53] Piscopiello M et al. P2X7 receptor is expressed in human vessels and might play a role in atherosclerosis. International Journal of Cardiology. 2013;**168**(3):2863-2866. DOI: 10.1016/j.ijcard.2013.03.084

[54] Tang J, Kern TS. Inflammation in diabetic retinopathy. Progress in Retinal and Eye Research. 2011;**30**(5):343-358. DOI: 10.1016/j.preteyeres.2011.05.002

[55] Frank RN. Diabetic retinopathy. Progress in Retinal and Eye Research. 1995;**14**(2):361-392. DOI: 10.1016/1350-9462(94)00011-4

[56] Yang LP et al. Baicalein reduces inflammatory process in a rodent model of diabetic retinopathy. Investigative Ophthalmology and Visual Science. 2009;**50**(5):2319-2327. DOI: 10.1167/iovs.08-2642

[57] Portillo JAC et al. Proinflammatory responses induced by CD40 in retinal endothelial and müller cells are inhibited by blocking CD40-traf2,3 or CD40-traf6 signaling. Investigative Ophthalmology and Visual Science. 2014;**55**(12):8590-8597. DOI: 10.1167/iovs.14-15340

[58] Portillo JAC et al. CD40 in retinal müller cells induces P2X7-dependent cytokine expression in macrophages/ microglia in diabetic mice and development of early experimental diabetic retinopathy. Diabetes. 2017;**66**(2):483-493. DOI: 10.2337/ db16-0051

[59] Liu X et al. The effect of A2A receptor antagonist on microglial activation in experimental glaucoma. Investigative Ophthalmology and Visual Science. 2016;**57**(3):776-786. DOI: 10.1167/iovs.15-18024

[60] Boia R, Ambrosio AF, Santiago AR. Therapeutic opportunities for caffeine and A2A receptor antagonists in retinal diseases. Ophthalmic Research. 2016;**55**(4):212-218. DOI: 10.1159/000443893

[61] Aires ID et al. Intravitreal injection of adenosine A2A receptor antagonist reduces neuroinflammation, vascular leakage and cell death in the retina of diabetic mice. Scientific Reports. 2019;**9**(1):1-14. DOI: 10.1038/ s41598-019-53627-y

[62] Ferrari D, Los M, Bauer MKA, Vandenabeele P, Wesselborg S, Schulze-Osthoff K. P2Z purinoreceptor ligation induces activation of caspases with distinct roles in apoptotic and necrotic alterations of cell death. FEBS Letters. 1999;**447**(1):71-75. DOI: 10.1016/ S0014-5793(99)00270-7

[63] McLarnon JG, Ryu JK, Walker DG, Choi HB. Upregulated expression of purinergic P2X7 receptor in Alzheimer disease and amyloid-β peptide-treated microglia and in peptide-injected rat hippocampus. Journal of Neuropathology and Experimental Neurology. 2006;**65**(11):1090-1097. DOI: 10.1097/01. jnen.0000240470.97295.d3

[64] Notomi S et al. Critical involvement of extracellular ATP acting on P2RX7 purinergic receptors in photoreceptor cell death. The American Journal of Pathology. 2011;**179**(6):2798-2809

[65] Caprioli J, Coleman AL. Blood pressure, perfusion pressure, and Glaucoma. American Journal of Ophthalmology. 2010;**149**(5):704-712. DOI: 10.1016/j.ajo.2010.01.018

[66] Osborne NN. Mitochondria: Their role in ganglion cell death and survival in primary open angle glaucoma. Experimental Eye Research. 2010;**90**(6):750-757. DOI: 10.1016/j. exer.2010.03.008

[67] Takenouchi T et al. P2X7 receptor signaling pathway as a therapeutic target for neurodegenerative diseases. Archivum Immunologiae et Therapiae Experimentalis (Warsz). 2010;**58**(2):91-96. DOI: 10.1007/s00005-010-0069-y

[68] Skaper SD, Debetto P, Giusti P. The P2X 7 purinergic receptor: From physiology to neurological disorders. The FASEB Journal. 2010;**24**(2):337-345. DOI: 10.1096/fj.09-138883

[69] Resta V et al. Acute retinal ganglion cell injury caused by intraocular pressure spikes is mediated by endogenous extracellular ATP. The European Journal of Neuroscience. 2007;**25**(9):2741-2754. DOI: 10.1111/j.1460-9568.2007.05528.x

[70] Hu H et al. Stimulation of the P2X7 receptor kills rat retinal ganglion cells in vivo. Experimental Eye Research. 2010;**91**(3):425-432. DOI: 10.1016/j. exer.2010.06.017

[71] Kuehn MH, Fingert JH, Kwon YH. Retinal ganglion cell death in glaucoma: Mechanisms and neuroprotective strategies. Ophthalmology Clinics of North America. 2005;**18**(3):383-395. DOI: 10.1016/j.ohc.2005.04.002

[72] David P, Lusky M, Teichberg VI. Involvement of excitatory neurotransmitters in the damage produced in chick embryo retinas by anoxia and extracellular high potassium. Experimental Eye Research. 1988;**46**(5):657-662. DOI: 10.1016/S0014-4835(88)80054-X

[73] Louzada-Junior P, Dias JJ, Santos WF, Lachat JJ, Bradford HF, Coutinho-Netto J. Glutamate release in experimental Ischaemia of the retina: An approach using microdialysis. Journal of Neurochemistry. 1992;**59**(1):358-363. DOI: 10.1111/j.1471-4159.1992.tb08912.x

[74] Zhang X, Li A, Ge J, Reigada D, Laties AM, Mitchell CH. Acute increase of intraocular pressure releases ATP into the anterior chamber. Experimental Eye Research. 2007;**85**(5):637-643. DOI: 10.1016/j.exer.2007.07.016

[75] Reigada D, Lu W, Zhang M, Mitchell CH. Elevated pressure triggers a physiological release of ATP from the retina: Possible role for pannexin hemichannels. Neuroscience. 2008;**157**(2):396-404. DOI: 10.1016/j.neuroscience.2008.08.036

[76] Sakamoto K et al. P2X7 receptor antagonists protect against N-methyl-d-aspartic acid-induced neuronal injury in the rat retina. European Journal of Pharmacology. 2015;**756**:52-58. DOI: 10.1016/j.ejphar.2015.03.008

[77] Soliño M et al. The expression of adenosine receptors changes throughout light induced retinal degeneration in the rat. Neuroscience Letters. 2018;**687**:259-267. DOI: 10.1016/j.neulet.2018.09.053

[78] Zhang QL, Wang W, Li J, Tian SY, Zhang TZ. Decreased miR-187 induces retinal ganglion cell apoptosis through upregulating SMAD7 in glaucoma. Biomedicine & Pharmacotherapy. 2015;**75**:19-25. DOI: 10.1016/j.biopha.2015.08.028

[79] Zhang QL et al. Down-regulated miR-187 promotes oxidative stress-induced retinal cell apoptosis through P2X7 receptor. International Journal of Biological Macromolecules. 2018;**120**:801-810

[80] Nishijima K et al. Vascular endothelial growth factor—A is a survival factor for retinal neurons and a critical neuroprotectant during the adaptive response to ischemic injury. The American Journal of Pathology. 2007;**171**(1):53-67. DOI: 10.2353/ajpath.2007.061237

[81] Fleck BW, McIntosh N. Pathogenesis of retinopathy of prematurity and possible preventive strategies. Early Human Development. 2008;**84**(2):83-88. DOI: 10.1016/j.earlhumdev.2007.11.008

[82] Cavallaro G et al. The pathophysiology of retinopathy of prematurity: An update of previous and recent knowledge. Acta Ophthalmologica. 2014;**92**(1):2-20

[83] Liu Xiao-Ling XL et al. Genetic inactivation of the adenosine A 2A receptor attenuates pathologic but not developmental angiogenesis in the mouse retina. Investigative Ophthalmology and Visual Science. 2010;**51**(12):6625-6632. DOI: 10.1167/iovs.09-4900

[84] Zhou R et al. Adenosine A2A receptor antagonists act at the hyperoxic phase to confer protection against retinopathy. Molecular Medicine. 2018;**24**(1):1-13. DOI: 10.1186/s10020-018-0038-1

[85] Steel D. Retinal detachment. BMJ Clinical Evidence. 2013;**2014**(September):1-32

[86] Gao S, Li N, Wang Y, Zhong Y, Shen X. Blockade of adenosine A2A receptor protects photoreceptors after retinal detachment by inhibiting inflammation and oxidative stress. Oxidative Medicine and Cellular Longevity. 2020:1-12. DOI: 10.1155/2020/7649080

[87] Osborne NN, Casson RJ, Wood JPM, Chidlow G, Graham M, Melena J. Retinal ischemia: Mechanisms of damage and potential therapeutic strategies. Progress in Retinal and Eye Research. 2004;**23**(1):91-147. DOI: 10.1016/j.preteyeres.2003.12.001

Chapter 6

Leukaemia: The Purinergic System and Small Extracellular Vesicles

Arinzechukwu Ude and Kelechi Okeke

Abstract

Haematopoiesis is a tightly regulated process, by intrinsic and extrinsic factors, to produce lifelong blood cell lineages within the bone marrow. In the bone marrow microenvironment, mesenchymal stem cells and haematopoietic stem cells play important roles to ensure that haematopoiesis is maintained. These cells contain purines and pyrimidines that control intercellular process such as energy transport. However, in some cases, this process may be misregulated thus leading to the production of various diseases, including leukaemia. As a result, bone marrow cells may be stimulated via stress or induced hypoxia, and this leads to the release of purine and pyrimidine nucleotides and nucleosides into the extracellular space, and activation of autocrine/paracrine feedback loops. These extracellular nucleotides and nucleosides, and their respective cell surface receptors are involved in purinergic signaling that control different physiologic functions in cells including proliferation, differentiation, and cell death. These extracellular nucleotides and nucleosides include ATP, UTP, adenosine diphosphate (ADP), UDP and adenosine however the most important players are ATP and its metabolite adenosine. ATP is degraded via a sequential activity of ectonucleotidases. ATP, adenosine and these ectonucleotidases play very important roles in the tumour microenvironment crucial to disease development, progression, and aggressiveness by modulating immune response to leukaemia treatment and increasing homing of leukaemic cells.

Keywords: cancer, communication, vesicles, transplantation

1. Introduction

Leukaemia is a malignant disorder involving early blood-forming cells of the bone marrow in which there is lack of control in the haematopoietic process. As a result, abnormal or immature blood cells accumulate in the bone marrow, bloodstream or lymphatic system thus causing debilitating effects in the patient [1, 2]. Leukaemia is a heterogenous disease hence its diagnosis is based on a complete blood workup (full blood count) and bone marrow studies (aspiration and biopsy) incorporating clinical, morphological, immunophenotypical, cytogenetic and genetic data.

Leukaemia is classified into different types depending on the blood cell type that the cancer originates. Leukaemia could be myeloid (myelogenous) if it involves the myeloid lineage such as red blood cells (RBC), platelets and white blood cells

(WBCs) or lymphoid (lymphocytic/lymphoblastic) if it involves cells originating from the lymphoid lineage such as lymphocytes [1, 3]. Leukaemia is also classified into acute and chronic based on the progression of the disease. Acute leukaemia is often abrupt and fast-growing due to the accumulation of immature cells in the bone marrow whilst chronic leukaemia is rather slow growing [1, 3]. Therefore, leukaemia is categorized into four main groups: acute myeloid leukemia (AML), acute lymphocytic leukaemia (ALL), chronic myeloid leukaemia (CML) and chronic lymphocytic leukaemia (CLL).

2. Epidemiology and aetiology of leukaemia

Leukaemia is one of the most common cancers in the world and accounts for about 2.6% of all cancer cases worldwide [4]. In 2020, leukaemia was reported to be the 13th most diagnosed cancer case and 10th leading cause of cancer death [5]. Caucasians especially the men are more predisposed to the disease compared to other ethnic groups. The disease progresses with age hence adults are more susceptible than children. However, the outlook for patients is much better than three decades ago, with better cure rates and longer-term disease-free survival. Chemotherapy and radiotherapy are the mainstay treatment for leukaemia however the overall survival rate has improved remarkably in recent years with almost half of the population diagnosed with leukaemia surviving for at least five years or more [3, 6]. This is due to the advent of different therapies including stem cell transplantation, targeted therapy, combined therapy, immune cell engineering such as chimeric antigen receptor (CAR)-T cell therapy, innovative clinical trials and immunotherapies, and patient's improved access to these therapies [6, 7].

Though an increase in age is a well-known factor, the aetiology of leukaemia remains unclear. Exposure to certain viruses (for example, human T-cell leukaemia virus; HTLV-1), ionizing radiation, chemicals such as benzene or formaldehyde, smoking and other environmental cues have all been reported to be risk factors for leukaemia [8–10]. These environmental cues induce genetic and chromosomal aberrations such as chromosomal deletions or translocations, point mutations and epigenetic factors, that cause maturation arrest of the normal haematopoietic stem cells (HSCs). This leads to development of leukaemia due to the proliferation and clonal expansion of leukaemic stem cells (LSC). Leukaemia could also arise following exposure to chemotherapy, radiotherapy and/or stem cell transplantation [6, 7].

3. Leukaemia and the purinergic system

Haematopoiesis is a tightly regulated process that leads to lifelong production of a sustainable pool of functional blood cells within a compartmentalized bone marrow. The bone marrow consists of osteocytes, osteoblasts, osteoclasts, the bone matrix, perivascular cells, immune cells, sinusoidal endothelium as well as regenerative cells; mesenchymal stem cells (MSC) and HSC that inhabit a unique hypoxic microenvironment [11, 12]. These cells are responsible for the provision of cell signals required for the support and regulation of haematopoiesis as well as maintenance of homeostasis within the bone marrow microenvironment [12, 13].

Additionally, there is prevalence of purines and pyrimidines in these bone marrow cells thus modulating intracellular processes such as energy transport [14, 15]. These cells can also release the purine and pyrimidine nucleotides and nucleosides into the extracellular space under normal circumstance in the absence of any stimulus. However, the stimulation of these bone marrow cells via stress or induced hypoxia leads to the release of purine and pyrimidine nucleotides and nucleosides, and activation of autocrine/paracrine feedback loops [12–15]. These extracellular nucleotides and nucleosides, and their respective cell surface receptors control a form of cell-to-cell communication called purinergic signaling. This complex network controls various physiologic functions in the body, including cell proliferation, differentiation, and cell death. As a result, any aberration to this network will lead to the development of diseases such as leukaemia [11, 16–19].

These extracellular nucleotides and nucleosides include ATP, UTP, adenosine diphosphate (ADP), UDP and adenosine however the most important players are ATP and its metabolite adenosine [11, 19, 20]. ATP is degraded via a sequential activity of four ectonucleotide enzyme subtypes (ectonucleotide pyrophosphate, ectonucleotide triphosphate diphosphohydrolase, alkaline phosphatase and ecto5-nucleotidase) leading to adenosine, which binds P1 receptors, as end-product [2, 11, 17, 18, 21–24]. Thus, extracellular ATP is the primary source of adenosine, and adenosine is degraded by adenosine deaminase (ADA). HSCs express several receptors for nucleotides and nucleosides, which belong to two different purinergic receptor families, P1 and P2. The P2 family includes eight receptors that have been identified so far (P2Y1, P2Y2, P2Y4, P2Y6, P2Y11, P2Y12, P2Y13, P2Y14), which are G protein-coupled receptors that respond to stimulation by ATP, ADP, UTP or UDP, depending on the receptor subtype and seven ionotropic receptors (P2X1, P2X2, P2X3, P2X4, P2X5, P2X6, P2X7), which are sensitive to ATP [15, 18, 25]. The P1 receptor family consists of four G protein-coupled receptor subtypes, A_1, A_{2A}, A_{2B}, and A_3, which are activated by adenosine.

Extracellular ATP is a crucial component of the tumour microenvironment. The balance between extracellular ATP and adenosine is pivotal to cancer progression, promoting proliferation of HSCs and thus, cancer cell proliferation [19]. Extracellular ATP also mediate proinflammatory activity on specific P2Y (P2Y1, P2Y2, P2Y4, P2Y6 and P2Y12) and P2X (P2X4 and P2X7) receptors. In addition, extracellular ATP exerts angiogenic effects on P2Y1, P2Y2 and P2X7 receptors and immunosuppressive effects on P2Y11 receptors thereby affecting interaction with the immune system via trafficking of granulocytes and monocytes and inhibiting proliferation and migration of leukemic cells [2, 16, 18, 23, 24, 26, 27]. Cellular stress or damage to stromal host cells promotes and controls inflammatory response resulting in ATP release and subsequent accumulation of ATP in extracellular space, which can be beneficial or harmful to the host depending on its concentration.

Notably, the concentration gradient of ATP is maintained by two important enzymes, CD39 (ectonucleotide triphosphate diphosphohydrolase-1) and CD73 (ecto5-nucleotidase) that are found in abundance in immune cells, endothelial cells, and tumour cells [18, 19, 28]. This enzymatic chain is also responsible for adenosine production, which can accumulate in the tumour environment and stroma where it acts as a potent immunosuppressant that exerts its effects mainly at A2A receptors. In addition, adenosine also modulates cell growth when acting on A3 receptors. Multiple mechanisms have been implicated in adenosine effects and these include deregulation of mononuclear phagocyte cell differentiation and maturation, suppression of effector T cells, inhibition of T helper 1 cell cytokine production, and generation of an angiogenic and matrix remodeling environment [17, 21, 23]. Extracellular adenosine also modulates and protects cells and tissues from excessive inflammatory and immune responses, which favor angiogenesis and thus, carcinogenesis [2, 18, 19, 29].

Adenosine acts on the A2A, A2B and A3 receptors to induce macrophages to promote the release of anti-inflammatory cytokine such as tumour necrosis factor (TNF), nitric oxide (NO), macrophage inflammatory protein (MIP), interleukin (IL)-6, IL-10, and IL-12. Therefore, ATP exerts dual mechanism in cancer, facilitating pro- and anti-tumoral immune response whilst adenosine is a known immunosuppressive mediator facilitating tumour immune evasion [11, 18, 19, 23]. An immunosuppressed microenvironment and inflammation enhance the development and metastasis of cancer via release of a wide variety of cytokines and other proinflammatory markers.

In various cancers, including leukaemia, it is widely acknowledged that purinergic receptors of the P2 family (P2Rs) are required for anti-cancer immune response induced by chemotherapy. In leukaemia, chemotherapeutic agents such as doxorubicin and daunorubicin induce intracellular ATP release to create an immunosuppressed tumour microenvironment, which leads to cell death [2, 19, 20]. Once released, the extracellular ATP can bind to P2 receptors to regulate cell invasion and migration. Treatment with ATP inhibitors or ATP analogues has a strong cytotoxic effect on several tumours, including leukaemia [23, 24]. This leads to a decrease in the intracellular concentration of ATP until it falls to undetectable levels when cells enter secondary necrosis. Low ATP doses have a growth-promoting effect and depending on the P2 receptor subtypes expressed, tumour cells may be more sensitive to the death inducing or the trophic effect of ATP [21].

Recent developments have shown that the purinergic system is potentially beneficial in leukaemia [2, 16, 29]. ATP/P2X7 axis is very vital in regulating the proliferation and homing of leukaemic initiating cells. Purinergic receptors, P2XRs, particularly P2X7R, are elevated in patients with acute and chronic forms of leukaemia [2, 11, 15, 16]. Studies show that patients with acute and chronic forms of leukaemia express higher levels of P2X7 than the healthy bone marrow controls, with the P2X7-loss-of-function polymorphism linked to increased capacity to evade apoptosis and therefore, to progression of chronic leukaemia [2, 11, 15, 16]. Knockdown studies have shown that ATP in the leukaemia microenvironment decreases upon knockdown of P2X7 thus leading to the lysis of leukaemic cells [16, 18, 19, 29]. P2X7 and P2Y11 were also identified in leukaemic cell lines, HL-60 and NB-4 cell lines [11, 29]. Leukaemic cell lines also express elevated levels of ecto-enzymes CD39 and CD73 [2, 19, 22]. In another study, the levels of these enzymes that are concerned with purine degradation, CD73, ADA and purine nucleoside phosphorylase were varied in patients with different forms of chronic leukaemia [19, 21, 22]. This infers that these enzymes may be beneficial to the survival of leukemic cells and promote metastatic spread.

Since promotion of inflammatory response is a hallmark of cancer, and ATP and adenosine exert pro-inflammatory and anti-inflammatory activities, the development of innovative agonists and antagonists that target these specific receptors will be a relevant therapeutic mechanism in leukaemia. Adenosine analogues have been proposed as a possible differentiation-inducing agent against AML. Adenosine exerts cytotoxic effects on leukaemic cells after ectoenzymic breakdown of ATP, with ATP increasing the cation permeability of acute and chronic leukaemia lymphocytes and ADP increasing the calcium mobilization of myeloid leukaemia cells [23, 24]. A voltage-gated potassium channel blocker, 4-aminopyridine, also induced apoptosis in human AML cells via increasing the calcium ions through P2X7 receptor pathways. Activation of the P2X7 receptor induces the formation of reactive oxygen species in erythroleukemia cells whilst P2X7 receptor agonists mediate cation uptake into human myeloid leukaemic cells [2, 23, 25, 29]. Evidence suggest that P2XR expression and activity may be relative to disease severity and depends on the level of activation as shown in lymphocytes of patients with varying forms of B-cell CLL [16, 23, 25]. Remission in patients with CML has also been associated with a frameshift polymorphism of the P2X5 receptor that elicits an allogeneic cytotoxic T lymphocyte response. Low levels of ATP triggered anti-proliferative effects in AML cells and except for P2X2, P2X3 and P2X6, AML cells are known to express all subtypes of P2Rs thereby suggesting targeting ATP is a promising alternative therapy in AML with favourable outcome in patients [2, 19, 21, 22]. Taken together, ATP/P2R axis demonstrate sustainable leukaemia-initiating cell signaling activities thus suggesting targeting and inhibiting ATP/P2XR signaling may completely hinder leukemogenesis.

4. The role of ectonucleotidases in leukaemia

Metabolic stress due to hypoxia, ischemia, and pro-inflammatory signals lead to abundant release of ATP in the extracellular space within the tumour microenvironment. CD39 and CD73 are ectonucleotidases that catabolize high extracellular levels of ATP to produce molecules that modulate intracellular calcium levels and activate the purinergic receptors [22, 28]. Under normal conditions, these ectonucleotidases potentially stimulate immune cells to fight against the tumour. However, these ectonucleotidases are also capable of modulating immune response to favour tumour growth.

CD39 converts extracellular ATP or ADP to adenosine monophosphate (AMP) whilst CD73 converts AMP to adenosine [22, 28]. Adenosine then accumulates in the tumour microenvironment thereby supporting tumour growth by skewing immune cells towards tolerance. Therefore, these enzymes have significant effects in pathogenesis and progression of leukaemia by enhancing chemoresistance and homing of leukaemic cells [22]. In support of this, high expression of ectonucleotidases increase homing of leukaemic cells to protected niches, survival, and proliferation of leukaemic cells and modulation of immune cells towards tolerance.

Since these enzymes are deeply involved in the pathogenesis of leukaemia, their expression in leukaemia may be of prognostic value in leukaemia, marking disease progression and aggressiveness, and immunosuppression. They may act as reliable markers to monitor and distinguish specific cellular populations or subset of patients characterized by a different prognosis. CD39 and CD73 are expressed in leukaemia of lymphoid and myeloid lineage such as AML, B-ALL and CLL [2, 19, 22]. In these subtypes of leukaemia, there's a significant crosstalk between these leukaemic cells and the bystander non-leukaemic cells. Therefore, the levels of these ectonucleotidases, both on the leukaemic and bystander non-leukaemic cells, may be used as disease markers or prognostic factors.

In ALL, CD73 was identified as a differentially expressed molecule in a genome-wide analysis, which compared 270 newly diagnosed ALL patients to normal B cell progenitors [30]. Following chemotherapy, flow cytometry validation revealed that these patients had upregulated expression of these enzymes thereby suggesting that CD73 could serve as a useful marker for minimal residual disease (MRD) and predicting ALL relapse following leukaemia treatment.

In AML, a more permissive immune environment is associated to high expression of CD39 and CD73, which favour leukaemia progression and aggressiveness [2, 19, 22]. AML cells release ATP following exposure to chemotherapy and upregulate CD39 expression on immune cells, and through this they modulate immune cells and skew them towards tolerance. In contrast, increased expression of CD73 on immune cells elicits a reawakening of the immune response. However, CD73 can also influence leukaemic cell proliferation and aggressiveness via CAAT enhancer-binding protein alpha (CEBPA) gene [19, 22].

In CLL, an increase in expression level of CD39 in patients is linked to an increase in circulating leukaemic cells as this contributes to disease progression and aggressiveness [2, 19, 22]. Thus, expression of CD39 on these immune cells can be used as a marker for advanced disease stage and a predictive factor of treatment requirement. An increased expression of CD73 is significant in CLL cells and is associated with a higher cellular turnover, an increased recirculation to and from the lymphoid niche, a more aggressive clinical behavior and associated with time to disease progression after chemotherapy [2, 22]. Therefore, CLL cells are well equipped with CD39/CD73 that protect these cells from drug-induced toxicity thereby causing MRD, a reservoir that fuels disease progression. In addition, adenosine, which is produced by these enzymes, further acts on immune cells and polarize them towards tolerance thus sustaining leukaemic cell expansion. Upregulation of CD73 and adenosine is mediated by hypoxia thus influencing the metabolic adaptation of immune cells surrounding the tumour. This induces a switch towards glycolysis, with upregulation of glucose and lactate transporters and of lactate dehydrogenase and pyruvate kinase [18]. This further links hypoxia and CD73/adenosine in a common axis in which the tumour microenvironment is reshaped with tumour-supportive and immunosuppressive features.

5. Leukaemia and small extracellular vesicles

Extracellular vesicles (EVs) are nanosized lipid bilayered membrane vesicles that are virtually released by all cells and secreted at higher numbers in cancer cells. They can also be found in varied body fluids such as semen, blood, follicular fluid, and urine. Based on their size, intercellular origin and release mechanisms, these vesicles are grouped into microvesicles (≥200 nm and ≤1000 nm), exosomes (≤200 nm) and apoptotic bodies [31, 32]. These vesicles originate from plasma membrane (microvesicles) and endosomes (exosomes) from inward budding of the plasma membrane into the cytoplasm to form multivesicular bodies (MVBs). These vesicles are then released into the extracellular milieu to be transported to neighbouring or distant cells to induce phenotypic and functional changes in the recipient cells [31, 33]. There are also other types of small EV that have been mentioned in recent literature such as exomeres, large oncosomes and enveloped viruses [34, 35]. Nevertheless, the detection and classification of these vesicles remain an uphill task due to this heterogeneity.

The recipient cells internalize their vesicles via different routes in different ways such as clathrin-dependent and clathrin-independent pathways like phagocytosis, macropinocytosis, caveolin-mediated uptake or lipid raft-mediated internalization [36, 37]. However, the mechanism of vesicular uptake and internalization remains a conundrum, but it is postulated that it depends on some factors found on the surface membrane of the recipient cells and the vesicles such as surface proteins, glycoproteins and glycolipoproteins. Therefore, it can be inferred that these surface membrane factors promote adhesion of the vesicles to the recipient cells thus facilitating their internalization.

Upon uptake, the function and fate of these vesicles differ depending on the physiological or pathological state of origin cells releasing and receiving these vesicles [36, 37]. These vesicles often carry a cargo of bioactive molecules, including proteins, lipids, and nucleic acids, which they can transfer to the recipient cells to modulate and reprogram the bone marrow microenvironment to promote their survival and induce biological effects in these cells [33, 38]. These vesicles play important roles in the regulation of immune stimulation or suppression that can drive inflammatory, autoimmune, and infectious disease pathology. EV could also alter the fate of their target cells by regulating gene expression through epigenetic changes in the recipient cells [36–38]. EV produced by cancer cells are in abundance in the tumour microenvironment and can enhance malignancy by transferring regulatory factors to normal cells within the tumour microenvironment. They also enhance anti-tumour immune response by inducing immunity to antigens that are carried by tumour EV.

In leukaemia, EV have a key role in the early stages of leukemogenesis as predominant EV population changes during leukaemia progression. Leukaemic cells utilize EV to transfer functional information to either MSC, HSC or myeloid progenitor cells in the BM microenvironment in amounts sufficient to induce phenotypic and functional changes in these cells, which are necessary for the development of leukaemia [33, 39]. These vesicles bridge the gap between leukaemic cells and the stromal cells that reside in the BM microenvironment thus initiating a crosstalk between these cells. This crosstalk between the leukaemic cells and stromal cells is crucial in remodelling and transforming the BM microenvironment into a leukaemia-permissive space, where leukaemic cells could proliferate, grow, and survive [29, 40, 41]. Leukaemia-derived EV transform and potentiate the phenotypic change of healthy MSC into cancer associated fibroblasts, which then proliferate, release inflammatory cytokines, and increase angiogenesis. As a result, leukemic cells survive and protect against apoptotic stimuli, including cytotoxic chemotherapy. AML-derived EV also alter the differentiation of stromal cells upon uptake thus leading to a decrease in the development of osteoblasts [42, 43].

The main mechanism through which EV promote the development and progression of leukaemia is not yet fully elucidated but much evidence supports through the delivery of microRNAs (miRNAs) into HSC within the BM microenvironment to change its characterization for developing into leukaemic cells [43–45]. These miRNAs are short non-coding RNAs that can mediate RNA silencing and regulate genes at the post-transcriptional level thus influencing the translocation of genes and signaling cascades. As thus, these small RNAs play vital roles in different cellular processes, including cell cycle, proliferation, angiogenesis, inflammation, immune reaction, and cell death [45, 46]. These small RNAs are either oncogenic or tumour suppressive, and once transferred from leukaemic cells to bone marrow microenvironment induce epigenetic changes that will support leukemogenesis. For example, in ALL, leukaemic cells release EV that carries miR-43a-5p to the bone marrow microenvironment and upon internalization, targets *Wnt* signaling pathway [45, 47]. This signaling pathway is vital for regulating haematopoiesis and induces the inhibition of osteogenesis in the bone marrow. Once suppressed, malignant HSC can then transform to leukaemic cells. MSC in the bone marrow microenvironment also secrete EV containing miR-21, which are then delivered into HSCs to enhance the development of B-ALL cells [40, 43]. These vesicle-derived miR-21 also interacts with transforming growth beta (TGF-β) to suppress the anti-tumour immune responses in the BM microenvironment.

In AML, the levels of serum-derived EV containing miR-10b, which is crucial in abrogating granulocytic/monocytic differentiation in HSCs, are elevated in patients compared to healthy individuals [12, 40, 43]. This suggests that miR-10b may play a vital role in inducing AML by enhancing the proliferative capacity of immature myeloid progenitors thus leading to the development of AML. Another miRNA, miR-4532 that targets the signal transducer and activator of transcription (*STAT-3*) signaling pathway, could also be transferred to HSCs from AML-derived EV to suppress the expression of leucine zipper downregulated in cancer 1 (*LDOC*), a well-known inhibitor of the *STAT-3* signaling pathway [47]. This leads to manipulation of the proliferative capacity of the cells in the BM microenvironment via this pathway. AML-derived EV also induce early leukemogenesis in myeloid progenitors through the transfer of miR-155 that inhibits *c-Myb* [31, 40, 47, 48]. *c-Myb* is a differentiating transcription factor in myeloid cells that induce differentiation arrest, which is critical in the development of AML. EV containing miR-155 and miR3-375 could also be transferred from MSC to AML cells to confer drug resistant phenotype against tyrosine kinase inhibitors [40]. Other miRNAs, such as miR-17-92 family, also induce chemoresistance in AML cells through activation of TGF-β and PI3K/Akt signalling axis. In CLL, EV released by leukaemic cells express a cell differentiating miRNA, miR-202-3p, which is taken up by stromal cells to influence the stroma cell transcriptome thereby resulting in altered growth characteristics [40].

However, miRNAs are not the only RNAs that have been implicated in the development and progression of leukaemia. Oncogenic messenger RNA (mRNA) derived from EV can be transferred to myeloid progenitors to encode nucleophosmin (*NPM1*) and FMS-like tyrosine kinase 3 *(FLT3)* gene with internal tandem duplication (ITD) (*FLT3-ITD*) thereby leading to the development of leukaemia [40]. Insulin-like growth factor 1 receptor (*IGF-1R*) and epidermal growth factor receptor (*EGFR*) mRNA enriched in AML vesicles are also transferred to stromal cells, leading to alterations in cell proliferation, secretion of growth factors and induction of downstream gene expression [31, 40, 47, 48]. AML-derived EV could also silence the expression of haematopoiesis-related growth factors such as *IGF-1,* C-X-C motif chemokine ligand 12 (*CXCL12), KIT* ligand and *IL-7*, and osteogenesis (osteocalcin; *OCN,* collagen type 1 alpha 1 chain; *Col1A1, IGF1*) whilst increasing the expression of genes that support AML growth such as Dickkop *wnt* signaling pathway inhibitor 1 (*DKK1*), *IL-6* and C-C motif chemokine ligand 3 (*CCL3*) [31, 40, 47, 48].

As thus, these enforce HSC committed to the myeloid progenitors. Anti-apoptotic proteins such as myeloid leukaemia 1 (*MCL-1*), B-cell lymphoma 2 (*BCL-2*) and B-cell lymphoma extra-large (*BCL-XL*) could also be transferred to the immature myeloid blasts to guarantee their survival in the BM microenvironment [46, 48]. By modulating the promoting loss of apoptosis or cell death, EV play a vital role in drug resistance. Drug-resistant AML cells also transfer p-glycoprotein to induce chemoresistance in drug-sensitive leukaemia cells via their nucleic acid and multidrug resistance protein 1 (*MRP1*) cargo [40].

In ALL, leukemic vesicles reduce mitochondrial respiration and cause metabolic switch, from oxidative phosphorylation to anaerobic glycolysis, in MSC thereby changing their ability to respond to metabolic changes [46]. This could provide the desired energy for ALL development in the BM microenvironment. In CLL, these vesicles also mediate the activation of *AKT* signaling pathway, which in turn induces the production of vascular endothelial growth factor (*VEGF*) to enhance progression of CLL [40]. Furthermore, EV from MSC of CLL patients could rescue leukaemic cells from drug-induced apoptosis and enhance their migratory capacity [31, 40, 49]. Lastly, CML-EV also promote the growth and maintenance of leukemic cells via internalization of their own EV or activation of *EGFR* signaling following delivery of amphiregulin protein [40, 49, 50]. Amphiregulin is a ligand of EGFR and stimulates cell growth, survival, and migration via juxtacrine, autocrine and paracrine signaling. EV from CML also contain *BCR-ABL* transcripts that induce increased proliferation in MSC upon uptake [31, 40].

Since leukaemic EV are enriched in tumour signature molecules and cargo antigens and immunological molecules associated with leukaemia cells, the role of EV within the leukaemic microenvironment may provide insight for therapeutic advances. Drug resistance is a substantial impediment to successful treatment in leukaemia. Leukaemia EV could act as circulating biomarkers for diagnosis and detection of leukaemia [31, 40, 46]. Their evaluation in body fluids especially peripheral blood and urine could provide relevant information for early and highly sensitive method for detection and monitoring of leukaemia progression in patients. EV could also be harnessed for gene delivery and personalized therapy in leukaemia [31, 50]. EV offer beneficial characteristics that synthetic vectors cannot such as high physiochemical stability, long distance communication, inherent cell signaling, cell-to-cell communication and bioactive delivery whilst protecting the interior cargo from stress-induced necrosis and the environment. EV can cross the blood brain barrier and target neuronal cells; this may be useful in treating central nervous system (CNS)-associated leukaemia, which has poor prognosis [46]. EV can also be synthetically modified by attaching cell-specific targeting ligands to the EV surface to cargo chemotherapeutic drugs directly to leukaemia cells thereby enhancing their functionality, specificity of cell targeting and decrease adverse immune response. Chemotherapeutic drugs such as imatinib and paclitaxel have been incorporated into EV and delivered to target IL-3 receptor on the CML blasts [40]. However, it is important to mention that the study of EV as biomarkers in clinical medicine is still a new field. No standard methods have been established yet for proper enrichment and isolation of these circulating vesicles. Varied methods such as differential ultracentrifugation, density gradient centrifugation, polymer-facilitated precipitation, immune-affinity isolation, and size exclusion chromatography are currently employed for isolation of EV. This heterogeneity in isolation techniques causes variability in the results thus affecting the purity of EV and subsequent precise molecular characterization of EV.

Author details

Arinzechukwu Ude[1*] and Kelechi Okeke[2]

1 University of the West of England Bristol, Bristol, UK

2 London South Bank University, London, UK

*Address all correspondence to: arinze_ude@yahoo.co.uk

References

[1] Bain BJ. Leukaemia Diagnosis. 5th ed. Wiley-Blackwell; 2017. p. 576. ISBN: 978-1-119-21054-2

[2] de Marchi E, Pegoraro A, Adinolfi E. P2X7 receptor in hematological malignancies. Frontiers in Cell and Developmental Biology. 2021;**9**:645605. DOI: 10.3389/fcell.2021.645605

[3] Cárceles-Álvarez A, Ortega-García JA, López-Hernández FA, Orozco-Llamas M, Espinosa López B, Tobarra-Sánchez E, et al. Spatial clustering of childhood leukaemia with the integration of the paediatric environmental history. Environmental Research. 2017;**156**:605-612. DOI: 10.1016/j.envres.2017.04.019

[4] IARC Working Group on the Evaluation of Carcinogenic Risk to Humans. Pharmaceuticals. Lyon (FR): International Agency for Research on Cancer; 2012 (IARC Monographs on the Evaluation of Carcinogenic Risks to Humans, No. 100A.) CHLORAMBUCIL. Available from: https://www.ncbi.nlm.nih.gov/books/NBK304324/

[5] Cancer Research UK Leukaemia Statistics. Available from: https://www.cancerresearchuk.org/health-professional/cancer-statistics/statistics-by-cancer-type/leukaemia#heading-Zero

[6] Wiseman DH. Donor cell leukemia: A review. Biology of Blood and Marrow Transplantation. 2011;**17**(6):771-789. DOI: 10.1016/j.bbmt.2010.10.010

[7] Joannides M, Grimwade D. Molecular biology of therapy-related leukaemias. Clinical and Translational Oncology. 2010;**12**(1):8-14. DOI: 10.1007/s12094-010-0460-5

[8] Bartenhagen C, Fischer U, Korn K, Pfister SM, Gombert M, Chen C, et al. Infection as a cause 237 of childhood leukemia: Virus detection employing whole genome sequencing. Haematologica. 2017;**102**(5):e179-e183. DOI: 10.3324/haematol.2016.155382

[9] Fircanis S, Merriam P, Khan N, Castillo JJ. The relation between cigarette smoking and risk of acute myeloid leukemia: An updated meta-analysis of epidemiological studies. American Journal of Hematology. 2014;**89**(8):125-132. DOI: 10.1002/ajh.23744

[10] Khalade A, Jaakkola MS, Pukkala E, Jaakkola JJ. Exposure to benzene at work and the risk of leukemia: A systematic review and meta-analysis. Environmental Health: A Global Access Science. 2010;**9**(31):1-8. DOI: 10.1186/1476-069X-9-31

[11] He X, Wan J, Yang X, Zhang X, Huang D, Li X, et al. Bone marrow niche ATP levels determine leukemia-initiating cell activity via P2X7 in leukemic models. Journal of Clinical Investigation. 2021;**131**(4):e140242. DOI: 10.1172/JCI140242

[12] Kumar B, Garcia M, Weng L, Jung X, Murakami JL, Hu X, et al. Acute myeloid leukemia transforms the bone marrow niche into a leukemia-permissive microenvironment through exosome secretion. Leukemia. 2017;**32**(3):575-587. DOI: 10.1038/leu.2017.259

[13] Kumar R, Godavarthy PS, Krause DS. The bone marrow microenvironment in health and disease at a glance. Journal of Cell Science. 2018;**131**(4):1-8. DOI: 10.1242/jcs.201707

[14] Kaebisch C, Schipper D, Babczyk P, Tobiasch E. The role of purinergic receptors in stem cell differentiation. Computational and Structural Biotechnology Journal. 2015;**13**:75-84. DOI: 10.1016/j.csbj.2014.11.003

[15] Huang Z, Xie N, Illes P, Di Virgilio F, Ulrich H, Semyanov A, et al. From purines to purinergic signalling: Molecular functions and human diseases. Signal Transduction and Targeted Therapy. 2021;**6**:162

[16] Zhang XJ, Zheng GG, Ma XT, Yang YH, Li G, Rao Q, et al. Expression of P2X7 in human hematopoietic cell lines and leukemia patients. Leukemia Research. 2004;**28**(12):1313-1322. DOI: 10.1016/j.leukres.2004.04.001

[17] Bagatini MD, dos Santos AA, Cardoso AM, Mânica A, Reschke CR, Carvalho FB. The impact of purinergic system enzymes on noncommunicable, neurological, and degenerative diseases. Journal of Immunology Research. 2018;**2018**:4892473. DOI: 10.1155/2018/4892473

[18] di Virgilio F. Purines, purinergic receptors, and cancer. Cancer Research. 2012;**72**(21):5441-5447. DOI: 10.1158/0008-5472.CAN-12-1600

[19] Feng LL, Cai YQ, Zhu MC, Xing LJ, Wang X. The yin and yang functions of extracellular ATP and adenosine in tumor immunity. Cancer Cell International. 2020;**20**(110):1-10. DOI: 10.1186/s12935-020-01195-x

[20] Salvestrini V, Zini R, Rossi L, Gulinelli S, Manfredini R, Bianchi E, et al. Purinergic signaling inhibits human acute myeloblastic leukemia cell proliferation, migration, and engraftment in immunodeficient mice. Blood. 2012;**119**(1):217-226. DOI: 10.1182/blood-2011-07-370775

[21] Antonioli L, Blandizzi C, Pacher P, Haskó G. The purinergic system as a pharmacological target for the treatment of immune-mediated inflammatory diseases. Pharmacological Reviews. 2019;**71**(3):345-382. DOI: 10.1124/pr.117.014878

[22] Vaisitti T, Arruga F, Giulia G, Deaglio S. Ectonucleotidases in blood malignancies: A tale of surface markers and therapeutic targets. Frontiers in Immunology. 2019;**10**(2301):1-17. DOI: 10.3389/fimmu.2019.02301

[23] Burnstock G. Pathophysiology and therapeutic potential of purinergic signaling. Pharmacological Reviews. 2006;**58**:58-86

[24] Burnstock G. Purinergic signalling. British Journal of Pharmacology. 2009;**147**:S172-S181

[25] Adinolfi E, Melchiorri L, Falzoni S, Chiozzi P, Morelli A, Tieghi A, et al. $P2X_7$ receptor expression in evolutive and indolent forms of chronic B lymphocytic leukemia. Blood. 2002;**99**:706-708

[26] Burnstock G. Purinergic signalling in the CNS. The Open Neuroscience Journal. 2010;**4**:24-30

[27] Burnstock G, Di Virgilio F. Purinergic signalling and cancer. Purinergic Signalling. 2013;**9**(4):491-540. DOI: 10.1007/s11302-013-9372-5

[28] Stagg J, Divisekera U, Duret H, Sparwasser T, Teng MW, Darcy PK, et al. CD73-deficient mice have increased antitumor immunity and are resistant to experimental metastasis. Cancer Research. 2011;**71**:2892-2900

[29] Conigrave AD, van der Weyden L, Holt L, Jiang L, Wilson P, Christopherson RI, et al. Extracellular ATP-dependent suppression of

proliferation and induction of differentiation of human HL-60 leukemia cells by distinct mechanisms. Biochemical Pharmacology. 2000;**60**:1585-1591

[30] Coustan-Smith E, Song G, Clark C, Key L, Liu P, Mehrpooya M, et al. New markers for minimal residual disease detection in acute lymphoblastic leukemia. Blood. 2011;**117**:6267-6276. DOI: 10.1182/blood-2010-12-324004

[31] Deng W, Wang L, Pan M, Zheng J. The regulatory role of exosomes in leukemia and their clinical significance. Journal of International Medical Research. 2020;**48**(8):1-10. DOI: 10.1177/0300060520950135

[32] Willms E, Cabañas C, Mäger I, Wood MJA, Vader P. Extracellular vesicle heterogeneity: Subpopulations, isolation techniques, and diverse functions in cancer progression. Frontiers in Immunology. 2018;**9**:00738. DOI: 10.3389/fimmu.2018.00738

[33] Borgovan T, Nwizu CC, Goldberg LR, Dooner MS, Wen S, Deltatto M, et al. Extracellular vesicles (EVS) shape the leukemic microenvironment. Blood. 2018;**132**(Supplement 1):5428. DOI: 10.1182/blood-2018-99-118955

[34] Lou G, Chen Z, Zheng M, Liu Y. Mesenchymal stem cell-derived exosomes as a new therapeutic strategy for liver diseases. Experimental & Molecular Medicine. 2017;**49**(6):e346

[35] Zhang J, Li S, Li L, Li M, Guo C, Yao J, et al. Exosome and exosomal microRNA: Trafficking, sorting, and function Genomics, Proteomics and Bioinformatics. 2015;**13**(1):17-24. DOI: 10.1016/j.gpb.2015.02.001

[36] Mulcahy LA, Pink RC, Carter DRF. Routes and mechanisms of extracellular vesicle uptake. Journal of Extracellular Vesicles. 2014;**3**(1). DOI: 10.3402/jev.v3.24641

[37] Durak-Kozica M, Baster Z, Kubat K, Stępień E. 3D visualization of extracellular vesicle uptake by endothelial cells. Cellular and Molecular Biology Letters. 2018;**23**(1). DOI: 10.1186/s11658-018-0123-z

[38] Butler JT, Abdelhamed S, Kurre P. Extracellular vesicles in the hematopoietic microenvironment. Haematologica. Ferrata Storti Foundation. 2018;**103**(3):382-394. DOI: 10.3324/haematol.2017.183335

[39] Cox MJ, Lucien F, Sakemura R, Boysen JC, Kim Y, Horvei P, et al. Leukaemia extracellular vesicles induce chimeric antigen receptor T cell dysfunction in chronic lymphocytic leukaemia. Molecular Therapy. 2021;**29**(4):1529-1540. DOI: 10.1016/j.ymthe.2020.12.033

[40] Pando A, Reagan JL, Quesenberry P, Fast LD. Extracellular vesicles in leukemia. Leukemia Research. 2018;**64**:52-60. DOI: 10.1016/j.leukres.2017.11.011

[41] Pando A, Fast L, Dubielecka PM, Chorzalska A, Wen S, Reagan J. Murine leukemia-derived extracellular vesicles elicit antitumor immune response. Journal of Blood Medicine. 2021;**12**:277-285. DOI: 10.2147/JBM.S308861

[42] Laurenzana I, Trino S, Lamorte D, Santodirocco M, Di Mauro L, Nazario B, et al. Role of acute myeloid leukemia derived-extracellular vesicles in dysregulating normal hematopoiesis. Blood. 2019;**134**(Supp 1.):1411. DOI: 10.1182/blood-2019-125544

[43] Lavoie JR, Rigg E, Munshi A, Mehic J, Muradia G, Rosu-Myles M. Characterization of leukaemia-derived

extracellular vesicles and subsequent effects on mesenchymal stem cell phenotype. Cytotherapy. 2018;**20**(5):SS2. DOI: 10.1016/j.jcyt.2018.02.050

[44] Huang D, Sun G, Hao X, He X, Zheng Z, Chen C, et al. ANGPTL2-containing small extracellular vesicles from vascular endothelial cells accelerate leukemia progression. Journal of Clinical Investigation. 2021;**131**(1):e138986. DOI: 10.1172/JCI138986

[45] Yan W, Song L, Wang H, Yang W, Hu L, Yang Y. Extracellular vesicles carrying miRNA-181b-5p affects the malignant progression of acute lymphoblastic leukemia. Journal of Translational Medicine. 2021;**19**(1):511. DOI: 10.1186/s12967-021-03174-w

[46] Izadirad M, Huang Z, Jafari F, Hamidieh AA, Gharehbaghian A, Li YD, et al. Extracellular vesicles in acute leukemia: A mesmerizing journey with a focus on transferred microRNAs. Frontiers in Cell and Developmental Biology. 2021;**9**:766371. DOI: 10.3389/fcell.2021.766371

[47] Raimondo S, Corrado C, Raimondi L, de Leo G, Alessandro R. Role of extracellular vesicles in hematological malignancies. BioMed Research International. 2015;**2015**:1-9. DOI: 10.1155/2015/821613

[48] Trino S, Lamorte D, Caivano A, de Luca L, Sgambato A, Laurenzana I. Clinical relevance of extracellular vesicles in hematological neoplasms: From liquid biopsy to cell biopsy. Leukemia. 2021;**35**(3):661-678. DOI: 10.1038/s41375-020-01104-1

[49] Bernardi S, Farina M. Exosomes and extracellular vesicles in myeloid neoplasia: The multiple and complex roles played by these “Magic Bullets”. Biology (Basel). 2021;**10**(2):105. DOI: 10.3390/biology100

[50] Hong CS, Jeong E, Boyiadzis M, Whiteside TL. Increased small extracellular vesicle secretion after chemotherapy via upregulation of cholesterol metabolism in acute myeloid leukaemia. Journal of Extracellular Vesicles. 2020;**9**(1):1800979. DOI: 10.1080/20013078.2020.1800979

Section 3

Future Perspectives

Chapter 7

Immunomodulatory Effects of a M2-Conditioned Medium (PRS® CK STORM): Theory on the Possible Complex Mechanism of Action through Anti-Inflammatory Modulation of the TLR System and the Purinergic System

Juan Pedro Lapuente

Abstract

Co-culture of primary or mesenchymal stem cells (MSC) with M2 macrophages produces a very special conditioned medium with a recognizable and stable cytokine pattern (PRS CK STORM), independent of the donor, with unique anti-inflammatory properties. This product can regulate certain pathways of inflammation in an anti-inflammatory manner, including TLR3, TLR4, the inflammasome, and the purinergic system. The anti-inflammatory action of PRS CK STORM is demonstrated both by its composition and by its action in *in vitro* and *in vivo* inflammatory models. The study of the mechanism of action showed changes in the pattern of toll-like receptors (TLR) and purinergic receptors, with an increase in the relative expression of mRNA encoding A2a and A3 receptors, together with a decrease in the relative expression of mRNA encoding P2X7 receptors. Second, it mitigated the adverse effects of a systemic inflammatory process in mice, especially in comparison with a known anti-inflammatory drug (Anakinra). Thus, due to its profile in terms of biosafety and efficacy, PRS CK STORM may be a strong candidate to treat inflammatory processes, such as cytokine storm associated with severe infectious processes, including COVID-19.

Keywords: co-culture, cytokines, ADP, cross-talk, toll-like receptors (TLRs), macrophages (M), mesenchymal stem cells (MSCs)

1. Introduction

Inflammation is the response of an organism's immune system to damage caused to its cells and vascularized tissues by bacterial pathogens and by any other biological, chemical, physical, or mechanical aggressor. Such an inflammatory response must be self-limiting in time and intensity since, if this is not the case and if there is no perfect coordination between the innate and adaptive immune systems, a severe systemic inflammatory syndrome with positive feedback systems will occur, eventually causing a cytokine storm that can lead to multi-organ failure [1, 2]. In the establishment, maintenance and termination of this cytokine storm, at the molecular level, in cases of sepsis and severe viral infections such as that associated with COVID-19, the toll-like receptors (TLRs), NOD-like receptors (NLRs), and RIG-like helicase receptors (RLRs), cytokines, chemokines and growth factors, and the purinergic system will be fundamental in the establishment, maintenance, and termination of this cytokine storm.

In the late 1990s, the ability of infectious agents (bacteria, viruses, zoonoses, or parasitic and/or fungal infections) to trigger cytokine storm syndrome (CSS) was first described with the recognition of a case series of hemophagocytic lymphohistiocytosis (HLH) of viral origin [3]. Such a cytokine storm is basically characterized by an exaggerated production of proinflammatory and profibrotic soluble mediators (especially IL-1β, IL-6, and TNF-α), together with an aberrant immunopathological reaction, involving an uncoordination between the innate and adaptive immunity system, there being generally an overactivation of the innate immune system, the main cellular actors being macrophages, dendritic cells, monocytes, neutrophils, and T lymphocytes [4–7]. As a consequence of this cytokine storm, a situation of multi-organ hyperinflammation will be provoked, which usually affects mainly the lung and pancreas, among other organs, and which usually results in acute respiratory distress syndrome (ARDS) and/or acute lung injury (ALI), which can lead to multi-organ failure.

Although the association of increased levels of proinflammatory and profibrotic cytokines and chemokines with increased levels of morbidity and mortality following an infectious process is well known, we still lack a suitable drug to treat the cytokine storm [8].

The innate immune system is able to recognize molecular structures specific to viruses, bacteria, fungi, and other pathogens; these structures are known as pathogen-associated molecular patterns (PAMPSs) [9–11]. PAMPSs are small-molecule sequences that are repeated in groups of pathogens recognized by the so-called pattern recognition receptors (PRRs). These include the toll-like receptors (TLRs) family of membrane receptors, NOD-like receptors (NLRs) and RIG-like helicase receptors (RLRs), among which the NLRP receptors stand out, oligomeric structures called inflammasomes, responsible for generating the mechanism of pyropoptosis by hyperproduction of hyperinflammatory cytokines, used as a trigger for the hyperproduction of IL-1β and IL-18 [12]. These molecular patterns are essential for the recognition of microorganisms by innate immunity cells, which respond differently depending on the microorganism identified [9–11].

Analyzing the detection capabilities of all these receptors, both DAMPS and PAMPS, we conclude that the main receptors involved in innate immunity against infections are TLR2, TLR3, TLR4, TLR7, TLR9, NOD1, NOD2, RIGI, and NLRP3. In **Table 1**, we summarize the PAMPS and DAMPS that are able to activate them [13–17].

Considering the different receptors involved in cytokine storms associated with infectious processes, we can deduce that the activation of the transcription factors AP-1 (activator protein 1) and NF-kβ (nuclear factor kappa light chain enhancer of activated B cells), both common denominators in almost all pattern recognition pathways, will provoke the dreaded cytokine storm, resulting in a state of generalized

	PAMPS		DAMPS		PAMPS		DAMPS
TLR2	SARS COV-2	++	Histonas extracelulares HMGB-1 HSP-5# Beta-defensinas Glycopeotides	**NOD1**	SARS COV-2	–	Metabolitos [100] DAP MDP
	Other viruses	–			Other viruses	+	
	Sepsis	++			Sepsis	–	
	Bacteria	++			Bacteria	++	
	Fungi	++			Fungi	–	
TLR3	SARS COV-2	+	Extracellular RNA	**NOD2**	SARS COV-2	—	Metabolitos [100] Fragmentos ARN
	Other viruses	++			Other viruses	++	
	Sepsis	–			Sepsis	–	
	Bacteria	–			Bacteria	++	
	Fungi	–			Fungi	–	
TLR4	SARS COV-2	++	Extracellular Histones Neutrophilic elastases HMGB-1 HSP-5# Fibronectina Fibrinogen Heparin sulfate Hyaluronan	**RIG1**	SARS COV-2	+	RNA fragments [101] DNA fragments miRNA
	Other viruses	++			Other viruses	+	
	Sepsis	++			Sepsis	–	
	Bacteria	++			Bacteria	–	
	Fungi	++			Fungi	–	
TLR7	SARS COV-2	++	Extracellular RNA miRNA DNA fragments	**NLRP3**	SARS COV-2	++	Extracellular hemoglobin Extracellular Histones Cholesterol crystals Proteins myeloid
	Other viruses	++			Other viruses	++	
	Sepsis	++			Sepsis	++	
	Bacteria	–			Bacteria	++	
	Fungi	–			Fungi	++	
TLR9	SARS COV-2	+	miRNA DNA fragments				
	Other viruses	++					
	Sepsis	–					
	Bacteria	++					
	Fungi	++					

TLR 3 receptors are activated mainly by viruses, although SARS COV-2 is not activated to the same extent as other viruses, such as influenza virus, respiratory syncytial virus (RSV), or herpes virus. TLR 4 receptors are activated mainly by bacteria, but activation is also seen in viruses, although it is unclear whether this is due to a posteriori DAMPS. TLR 7 receptors are mainly activated by viruses. TLR 9 receptors are activated by both RNA and DNA so that viruses, bacteria, fungi, or any pathogen can activate them. Cytoplasmic receptors, since they detect most pathogen fragments, including DNA and RNA, can be activated by both viruses and bacteria, although NOD1 is activated more strongly by bacteria. It appears that RIG I is only activated by viral and not bacterial RNA [13–17]. (++ maximal activation, + intermediate or mild activation, no activation).

Table 1.
TLR 2 receptors are activated mainly by bacteria and fungi, and although their activation has been described in SARS COV-2 infection, it is possible that this is due not directly to the SARS COV-2 virus but to the existence of a concomitant bacterial infection.

hyperinflammation. The most affected organs are lung or pancreas, with the consequent associated fibrotic reaction, producing irreparable anatomopathological damage with loss of function in the most affected organs.

A fact especially associated with the cytokine storm associated with SARS COV-2 is that a decrease in the production of type I interferons is observed, which causes dysregulation in the coordination of the innate and adaptive immunity systems, facilitating the appearance of the dreaded severe pneumonia that on many occasions determines the patient's admission to the ICU [18].

It is very difficult to explain the existence of a cytokine storm by the activation of a single receptor. If this were so, treatment of the cytokine storm by a single monoclonal antibody, for example, a monoclonal antibody against IL-1β or against IL-6, would always be effective, and this we know almost never occurs. Moreover, even if in the first instance only one of the receptors is activated, the simple initiation of its metabolic cascade will provoke the appearance of DAMPS that will stimulate other receptors. If we add to this the fact that in the majority of cytokine storms associated with infections we do not see a single causative pathogen, but rather a group of them, we will understand that there is almost always a joint activation of several of these receptors, producing between them phenomena of agonism and synergy, as well as antagonism [19]. Any of them can have agonistic relationships with others, if they are stimulated at the same time. However, if these same receptors are activated with a significant time lag between them of hours or even days, the most likely mutual relationship they will establish will be one of antagonism [19]. Thus, the types of cytokines and chemokines that will be released as a result of the activation of the different receptors will depend on the sets of receptors that are primarily activated by PAMPS and, once initiated, such release of pro-inflammatory and profibrotic mediators will be prolonged and augmented over time by positive feedback from the same receptors or even the addition of others, by the stimuli elicited by DAMPS, which could lead to reactive phenomena even autoimmunity.

In the cytokine storm, we must also consider the intervention of the purinergic system [20–22]. Extracellular adenosine triphosphate (eATP) or its enzymatic degradation products, such as ADP, AMP, and adenosine, can stimulate a number of membrane receptors [23]. More specifically, the P2X7 receptor is widely distributed on innate cells of the adaptive immune system, a system that constitutes the first line of defense against invading pathogens. These cells are lymphocytes, granulocytes, macrophages (including microglia), and dendritic cells in peripheral tissues [24–26]. Activation of the P2X7 receptor has been associated with the establishment and prolongation of inflammation and cytokine storm in septic infections, including SARS-COV-2 infection [27–29]. The stimulation of the P2X7 receptor by adenosine triphosphate (ATP) causes the activation of the NLRP3 inflammasome, and consequently of caspase 1, stimulating this the exaggerated secretion of IL-1β and IL-18 [30]. For all these reasons, the ideal immunomodulatory treatment of the cytokine storm associated with moderate and severe infections should include the P2X7 receptor (generating antagonism) or P1-like receptors (generating agonism) as a therapeutic target [29].

The treatments tested to date to control cytokine storms associated with infectious processes have been based on the use of monoclonal antibodies used alone or in combination. The hypothesis put forward by our group proposes as a treatment a biological therapy based on the use of allogeneic-conditioned medium derived from M2-type macrophages and enriched with mesenchymal stem cells (MSCs). Mesenchymal stem cells, placed in co-culture with macrophages, not only respond to macrophages and adjust their secretome accordingly but also induce macrophages to respond to them, creating a feedback loop that contributes to immune regulation [31]. In the complex composition of this conditioned medium are present all growth factors, cytokines, and chemokines that are naturally produced by M2-type macrophages and MSCs, associated with innate immunity respecting natural pleiotropic relationships. The immunomodulatory cytokine profile of this medium confers a potent anti-inflammatory and anti-fibrotic action, and even, thanks to the results obtained with the secretome of MSCs on macrophages stimulated with TLR7/8 ligand, possibly antiviral [32]. Different studies have shown that the secretome of these two cell types is modified and modulated when co-cultures of these cells are performed [33]. The immunomodulation mechanisms mediated by MSCs are due, among other factors, to the release of PGE-2 (prostaglandin E-2) and TSG-6 (TNF-stimulated gene 6 protein) [34]. M2 (anti-inflammatory) macrophages secrete high levels of IL-10 and low levels of IL-12p70 and IL-17, in a process that is directly mediated by other factors produced by MSCs (such as IL-6 and HGF) [35, 36]. It has been experimentally demonstrated that factors secreted by pro- or anti-inflammatory macrophages activate the immunomodulatory potential of MSCs. In this regard, IL-10 release by anti-inflammatory macrophages activates MSCs to release PGE-2 [37], which in turn modulates macrophages producing a cascade of additive molecular interactions in favor of immunotolerant and anti-inflammatory mechanisms. Basically, the polarization of macrophages to M2 type with immunoregulatory phenotype will be maintained [38], which, in turn, will express more IL-6, IL-10, and IGF-1 and inhibit their production of IL-12 and TNF-α. MSCs are also capable, through secretion of these same factors, of inhibiting the migration, maturation, and differentiation of dendritic cells [39–41]. Similarly, monocyte-derived M2 macrophages co-cultured with MSCs have been shown to increase mitochondrial function and ATP turnover, both in vitro and in vivo, resulting in an increase in the ADP/ATP ratio [42]. In addition, MSCs maintain ATPase and CD73 enzymatic activities on their surface, converting ATP to ADP and AMP to adenosine, respectively [43]. Adenosine, the last molecule in these reactions, has immunoregulatory functions through the activation of the P1 receptor [44]. Importantly, activation of monocyte P1 receptors, such as A2A and A2B, inhibits TNF-α production [44].

Several previous experiences demonstrate how secretomes from both cell types possess immunomodulatory properties. For example, direct injection of the supernatant of cultured mesenchymal stem cells (MSCs) containing a variety of growth factors, prostaglandins, and cytokines can be applied to the treatment of kidney injury [45]. Both co-culture with M2 macrophages and treatment with M2 macrophage supernatant have also been shown to increase endothelial cell viability in a bacterial lipopolysaccharide-generated lung sepsis model [46]. In addition, the efficacy and safety of multiple sclerosis treatment by intravenous infusion of conditioned medium from mesenchymal stem cell culture have also been demonstrated [47].

The advantage of using the complete conditioned medium versus one of its purified components lies in the synergistic mechanism between its different components [48], the result of subjecting the cell populations to a culture that, in vitro, attempts to emulate the anti-inflammatory, anti-fibrotic, and regenerative immunomodulatory microenvironment that occurs in vivo in diseased tissue.

2. Material and methods

Production and characterization of allogeneic-conditioned medium derived from M2-type macrophages and enriched with MSCs.

First, to obtain MSCs, a lipoaspirate sample was obtained from which the stromal vascular fraction (SVF) was extracted, following the protocol described by Lapuente et al. [49]. SVF was harvested by centrifugation under the same conditions as earlier-mentioned, seeded at a density of approximately 30,000 cells per cm^2 in 100-mm diameter culture plates (this and all culture plastic used was from Corning, Corning, NY, USA) and cultured at 37°C and 5% CO_2 in culture medium (DMEM + 10% fetal bovine serum (FBS) + 1% P/S). At 24 h, the culture was washed with phosphate-buffered saline (PBS) to remove nonadherent cells and the adherent cell population, called processed lipoaspirate (PLA), was cultured to subconfluence under the same conditions as earlier-mentioned, changing the culture medium three times a week and performing the necessary passages with trypsin 0.05% (Gibco), until a homogeneous population of mesenchymal-type stromal cells, also called mesenchymal stem cells (MSCs), was obtained. After culture, the cells were frozen at a freezing ramp of −0.5°C/min to −80°C in freezing medium composed of 10% dimethyl sulfoxide (DMSO, Sigma) in FBS or culture medium, then immersed in liquid N_2 and maintained until use.

Secondly, monocytes were isolated from one altruistic blood donation bag of 450 ml with 12% citrate–phosphate-dextrose (Grifols, Barcelona, Spain) from the blood bank of the Fuenlabrada Universitary Hospital. To isolate the leukocytes, each bag was divided into 50-ml tubes (Corning) and centrifuged at 1500 x g for 10 min at room temperature (RT). The intermediate band, leukocyte buffy coat, was collected and deposited on a clean tube. Immediately, 24 ml of this concentrate was carefully placed on 18 ml of Ficoll Histopaque 1077 (Sigma) and centrifuged at 400 × g for 30 min at room temperature (RT) and without brake. The mononuclear cell band was collected, and after adding PBS in a 1:1 ratio, centrifuged at 300 x g for 5 min at RT. The supernatant was discarded and the resulting pellet was resuspended in a fivefold volume of erythrocyte lysis buffer and incubated at RT for 10 min. Subsequently, a 10-fold volume of PBS was added and centrifuged under the same conditions as earlier-mentioned to obtain the cell pellet after discarding the supernatant. This last wash was repeated once more and, after this, the resulting peripheral blood mononuclear cell pellet (PBMC) was resuspended in CTS-AIM-V medium (Gibco) supplemented with 0.1% Dipeptiven 200 mg/ml (Frenesius Kabi Austria GmbH, Graz, Austria) and cultured in T-175 culture flasks (Corning, approximately 200 million PBMCs per flask) at 37°C and 5% CO_2 atmosphere for 90 min. The next step was to wash the flasks twice with plenty of PBS to remove unattached cells. The cells were immediately lifted with a cell scraper (Corning) to obtain a cell suspension in PBS, which was centrifuged for 5 min at 300 × g at room temperature. The resulting pellet was resuspended in AIMV + Dipeptiven + 10 ng/ml M-CSF (R&D Systems, McKinley, Minneapolis, USA) for co-culture. All cell counts and viabilities (trypan blue exclusion method) were performed with an automatic counter TC20 (BioRad, Hercules, CA, USA), strictly following the manufacturer's instructions, marking a lower threshold of 8 μm to disregard possible erythrocytes, platelets, and other contaminating cellular debris.

Third, co-culture was established to produce the conditioned medium. For this purpose, the obtained monocytes were seeded at a density of 500,000 cells/cm^2 in inserts (Transwels, with a polyethersulfone membrane of 1 μm pore size, from Corning) of 6-well plates and cultured under standard conditions for 4 days with the described medium. When the culture medium was removed, the inserts were washed twice with PBS and added to the plates on which the MSCs had been seeded and cultured in pass 4 (24 h earlier, in Corning 6-well plates at a density of 10,000 cells/cm^2 under standard conditions), previously washed twice with copious PBS and using CTS-AIMV-V medium supplemented with 0.1% Dipeptiven to maintain the co-culture under standard conditions of temperature and CO_2 concentration. The co-culture was maintained for 4 weeks by collecting the conditioned medium and adding fresh medium twice a week. To preserve the different collections, they were immediately frozen by immersion in carbonic snow and kept at −80°C until analysis. To perform the analysis, the cultures were thawed at 4°C and analyzed immediately after filtering through a 0.45-μm pore nitrocellulose filter (Merck KGaA, Darmstadt, Germany). To obtain secretome controls for MSCs and different monocytes, the different cell populations were cultured separately under the conditions described for co-culture.

Subsequently, phenotypic characterization of MSCs and monocytes was performed and samples were taken from both populations at time 0, 7 days, 14 days, and 28 days. MSCs were lifted with trypsin and monocytes with scraper as described earlier and, after centrifugation at 300 × g 5 min at 4°C, resuspended each cell type in PBS, permeabilized the monocytes with Perm/Wash buffer (BD Biosciences, Franklin Lakes, NJ, USA), and incubated the cells for 15 min at RT and in the dark with the following fluorochrome-conjugated antibodies (and their related isotypes as negative controls) at 1:50 concentration: CD73-APC, CD90-APC, CD45-FITC, HLA-DR-FITC, CD31-PE, CD68-FITC, and CD163-PE (all from BD Biosciences). The fluorescence minus one technique was used to adjust the voltages and compensate for fluorescence, and propidium iodide (Sigma) was used to determine dead cells according to the manufacturer's instructions. A Guava EasyCyte flow cytometer (Merck) was used to acquire the samples and InCyte software (Merck) was used to analyze the results.

To quantify the secretome of both cell types and the co-culture, 30 growth factors, cytokines, and chemokines were quantified using either ELISA or Multiplex assay (ProcartaPlex 23 PLEX, Invitrogen, Grand Island, NY, USA), strictly following the manufacturer's instructions. A Luminex Labscan 100 plate reader (Luminex Corporation, Austin, TX, USA) was used for the determinations. The molecules quantified by Multiplex were the following: MIP1-α, IL-2, IL-6, TIMP-1, IL-8, IL-10, IL-12 P70, IL-1 RA, RANTES, GM-CSF, leptin, HGF, MMP-3, MCP1, BNGF, EGF, adiponectin, TNF-α, MMP-1, TRAIL, FGF-2, PDGF-BB, and VEGF-A. For quantification of IGF-1, BMP-6, IL-1β, IL-4, TGF-β1, TGF-β3, and VEGF-C, a double sandwich ELISA technique was used following the manufacturer's instructions (DuoSet ELISA kit, R&D) and quantification was determined using an iMark plate reader (BioRad).

2.1 Generation of the *in vitro* inflammation model

THP-1 cell line culture and subsequent differentiation to macrophages were performed to generate in vitro models of biosafety and efficacy. THP-1 monocytic cells (CellLineService, cat. No.: 300356) were cultured and expanded using RPMI 1640 (Lonza, Basile, Switzerland) supplemented with 10% fetal bovine serum (FBS) (Corning, NY, USA), 1% penicillin/streptomycin (P/S) (Lonza), 1 mM sodium pyruvate (Lonza), and 1% MEM nonessential amino acids (Gibco, Thermo Fisher Scientific, Waltham, MA, USA) starting now THP-1 medium. Cells were maintained at a density ranging from 2.5×10^5 to 10^6 cells/ml to ensure adequate growth and a stable phenotype. Forty-eight hours prior to lipopolysaccharide (LPS) stimuli, cells were differentiated into resting macrophages using phorbol 12-myristate 13-acetate (PMA) (Sigma–Aldrich, Saint Louis, MO, USA) at 5 ng/ml in THP-1 medium as described in the protocol used by Park et al. [50]. After this differentiation process, the cells were used for our experiments. All cell cultures were maintained at 37°C in an atmosphere of 5% CO_2 and 98% relative humidity. The in vitro inflammation model was generated by differentiating 400,000 THP-1/ml cells in exponential growth phase into resting macrophages in 12-well plates (Nunc, ThermoFisher) (final volume 1 ml) and after 48-h pretreatment with PMA, the cells were washed three times with 0.5 ml of tempered THP-1 medium without PMA and allowed to incubate for 30 min before LPS stimuli. Once at rest, rest, the cells were treated with 10 ng/ml LPS (Sigma–Aldrich) in RPMI 1640 medium and the investigational product, which had been previously quenched at room temperature or quenched THP-1 medium, using as control the same THP-1 culture treated with the same amount of LPS, but adding in this case 10 μg/ml hydrocortisone. The final volume of each well was 1 ml with 400,000 cells each. Stimulation was carried out for 5 h.

2.2 Evaluation of the possible mechanism of action of the proposed conditioned medium

After 5 h of stimulation, supernatants were removed from each well, divided into aliquots, and flash-frozen by immersion in dry ice for further analysis. Total RNA was extracted from the cells using an RNeasy Plus Mini kit (Qiagen, Hilden, Germany), and extraction was carried out strictly according to the manufacturer's instructions. This kit included a genomic DNA removal step. The resulting RNA was eluted from the columns using nuclease-free water, divided into aliquots, and stored at −80°C to avoid degradation by environmental RNAases. From the 40 μl of RNA solution from each sample, an aliquot was extracted to assess RNA integrity and concentration. Total RNA integrity was assessed by agarose gel electrophoretic run of total RNA on a 2% agarose-TBE gel for 30 min at 120 V and 400 mAh. Quantification of total RNA was performed by a fluorimetric method using a highly sensitive fluorimetric kit (Qubit HS RNA quantification kit, Applied Biosystems, ThermoFisher). The cDNA was synthesized from total RNA for quantitative PCR of our genes of interest. A high-capacity cDNA reverse transcription kit (Applied Biosystems) was used for synthesis, and a total of 150-ng total RNA was used, for each synthesis reaction. Each sample had 5 cDNA synthesis reactions to achieve sufficient volume for downstream applications. The synthesis protocol was performed using an RNAase inhibitor following the manufacturers' recommendations, and their protocol was strictly followed. Random hexamers were used to perform reverse transcription of all mRNAs into double-stranded cDNA. After synthesis, the cDNAs were divided into aliquots and stored at −20°C, for later use. An aliquot of these cDNAs was extracted for quantification using a Qubit dsDNA HS Assay kit (Applied Biosystems). Primer concentrations were optimized using a cDNA pool to determine the most appropriate concentrations of the primers in the qPCR protocol. For such determination, a standard PCR was performed using a 2× PCR MasterMix (DreamTaq HotStart PCR MasterMix) (Applied Biosystems). Cycling conditions were 98°C for 3 min, then 35 cycles at 95°C for 45 s, 60°C for 30 s, and 72°C for 30 s. After these 35 cycles, the temperature was set at 72°C for 5 min and then held at 4°C indefinitely. The optimal primer concentration was determined by selecting the sharpest specific bands on agarose electrophoresis, uncontaminated by the presence of primer dimers at the front of the gel or nonspecific products. Ideal primer concentrations were 250 nM for forward and reverse primers. Primer sequences and amplicon sizes are attached in **Table 2**.

The following Thermo Fisher primers (coupled to FAM) were also used for A2a (Hs00169123_m1), A3 (Hs04194761_s1), and P2X7 (Hs00175721_m1) receptors.

Subsequently to perform qPCR, total RNA was extracted from 400,000 THP-1 cells using a Qiagen RNeasy plus mini kit (Qiagen, Hilden, Germany). THP-1 cells had been previously differentiated to resting macrophages using 5 ng/ml phorbol 12-

Gene name	Forward 5′→3′	Reverse 3′→5′	Gene ID	Amplicon size (bp)
TLR-2	CCACCTGCCTGGAACTCAG	CAGTCACCTGAGAGAACGCC	7097	216
TLR-3	AACGACTGATGCTCCGAAGG	CAGGGTTTGCGTGTTTCCAG	7098	207
TLR-4	TAGCGAGCCACGCATTCACA	TTAGGACCACCTCCACGCAG	7099	165
TLR-7	CCCTGGCCACAGACAATCAT	TCCTGTGACAGACGTTGGTG	51,284	210
TRAF6	CCGCGCACTAGAACGAGCAA	GGCAGTTCCACCCACACTATC	7189	157
MyD88	ATGCCTTCATCTGCTATTGCCC	GGCCTTCTAGCCAACCTCTTT	4615	175
RIPK1	ACTAGGTGGCAGGGTACAG	TGATCATGAGTCCCTGGGTT	8737	202
IKKβ	TGGACGTGGTCACAGACGGA	CGAGGAACCACCATGTGAGA	3551	203
NF-kβ	TTAGGAGGGAGAGCCCAC	AGTCGGATCTGTGGTTGAAA	4790	241
CASP1	CAGTCACACAAGAAGGGAGG	CCCCTTTCGGAATAACGGAC	834	227
NLRP3	GCTGGCATCTGGATGAGGAA	AAAGTTCTCCTGTTGGCTCG	114,548	247
GAPDH	GCACCACCAACTGCTTAGC	GCATGGACCGTGGTCATGAG	2597	131

Table 2.
Primer sequences and amplicon sizes.

mystirate 13-acetate (PMA) (Sigma–Aldrich, Saint Louis, MA, USA) in cell culture medium 48 h prior to the experimental model. A total of 0.75 μg of RNA was retrotranscribed to cDNA using a high-capacity cDNA synthesis kit (ThermoFisherScientific, Waltham, USA) employing random hexamers. For qPCR, 10 ng of cDNA per reaction was amplified using a Power SYBR-Green PCR master mix (Applied Biosystems, Thermo Fisher) on a StepOnePlus real-time PCR machine (Perkin Elmer, Waltham, MA, USA) and using the primers listed in **Table 1** (primer table). Thermal cycling conditions included an initial denaturation step at 95°C for 5 min, followed by 40 cycles at 95°C for 30 s, 60°C for 30s, and 72°C for 30s. Melting curve analysis of each qPCR was performed on the final products. Messenger RNA fold changes were calculated using the ΔΔΔCt method with GAPDH as a calibrator gene.

Finally, in order to have an approximation of the mechanism of action of our PRS® CK STORM conditioned medium, two studies were performed. Firstly, a quantification of the ATP/ADP ratio contained in the drug. The Sigma–Aldrich colorimetric ADP/ATP ratio assay kit (Ref: MAK135) was used for this purpose, and the Biorad iMark plate reader was used for its reading. Secondly, a quantification of extracellular ATP in THP-1 cells placed in culture, comparing the results when stimulated by LPS and/or treated with PRS® CK STORM conditioned medium. For this purpose, the ELISA ATP Assay Kit Colorimetric (Ref: ab83355) from Abcam was used, and the Biorad iMark plate reader was used for reading.

2.3 Generation of the *in vivo* inflammation model

To perform the experimental model of acute lung injury, the experimental model described by Stephens et al. [51] was used. For this model, 8–10 weeks old male C57BL/6 mice were used and administered 5 mg/kg of bacterial lipopolysaccharide (LPS) in 50 μl of physiological solution retro-orbital under anesthesia. To decrease the possible suffering of the mice due to LPS, they were administered buprenorphine hydrochloride in water at the established dose of 0.056 mg/ml. A total of 25 animals were used, with a number of animals per group of 5. The mice were conditioned 1 week prior to the procedure and were housed in standard conditions with access to food and water ad libitum with 12-h light/dark cycles at a temperature of 25°C and humidity greater than 40% over the course of the project.

- Basal group – animals given no LPS or treatment/basal data (n = 5; mice number 1–5).
- Vehicle control – animals administered 50 µl of vehicle (n = 5; mice number 6–10).
- Positive control group – animals given LPS but no treatment, vehicle only (n = 5; mice number 11–15).
- Gold standard-treated animals – LPS-treated animals given gold standard – Anakinra treatment at a dose of 200 mg/kg/day orally (defined in Stephens et al. [51]). (n=5; mice number 16–20).
- PRS® CK STORM-treated animals – LPS-treated animals given the drug intravenously (n = 5; mice number 21–25).

The vehicle used in the vehicle control group is PBS (phosphate-buffered saline). The gold standard treatment consists of the administration of Anakinra (IL-1Ra) before and after inoculation with LPS. The test group received the established dose of imatinib 24 h prior to LPS administration. The animals were administered the drug every 24 h starting from the retro-orbital administration of LPS. Treatment was administered intravenously (40 µl), upon generation of the model and then every 24 h thereafter. Blood samples were taken 24, 48, and 72 h after generation of the model from the submaxillary sinus from which 120 µl were collected for hemodynamic and general biochemical study, which was performed with the Comprehensive Diagnostic Profile protocol (#500–0038), of the VetScan V2 device (Abaxis). After 72 h from the generation of the model, the animals were sacrificed and exsanguinated and plasma samples were collected for final biochemical and cytokine analysis. Specifically, TNF-α, IL-1β, and IL-6 as proinflammatory cytokines and IL-10 as anti-inflammatory cytokine were quantified. They were analyzed by Luminex: MILLIPLEX MAP Mouse Cytokine/Chemokine Magnetic Bead Panel – Immunology Multiplex Assay (Cat: MCYTOMAG-70 K) (Merck). During the study, hyperthermia/hypothermia, respiratory distress, weight loss, food and water consumption, as well as the existence of other behavioral disorders were monitored. After euthanasia and subsequent necropsy, the major organs (heart, lung, liver, kidney, and spleen) were removed and fixed and preserved in 4% formalin for histological study. A small portion of each organ, prior to fixation, was preserved directly by immersion in dry ice to study inflammatory cytokine content in future assays. All animals underwent Irwin's test every 24 h to obtain neuropharmacology data following the protocol of Mathiasen et al. [52].

2.4 Statistical analysis

The MTT and cytokine release assays, as well as the cytokine analysis of the culture supernatants, the biochemical values of the blood of the different groups of treated mice, and the values obtained in the qPCRs, were subjected to statistical analysis. For these, a two-tailed Student's t-test was performed to obtain the p-value between the different experimental groups and to analyze the existence of significant differences ($p < 0.05$). Statistical analysis was performed with Excel (Microsoft, Albuquerque, NM, USA).

3. Results

3.1 Isolation of MSCs and monocytes

The yield achieved in the isolation of MSCs was approximately 1×10^5 cells per ml of lipoaspirate, and it was necessary to incubate for 16 days under the conditions described in the previous section to bring the culture to pass 4 (**Figure 1a**). The yield provided by the monocyte donor can be seen in **Table 3**.

Optical microscopy showed that the cell morphology of the adherent cells in the cultures corresponded to that of monocytes/macrophages.

The results of flow cytometric characterization of MSCs and monocytes from three co-cultures at the times studied are shown in **Figure 1**. Phenotypic characterization of MSCs shows the classic phenotype of CD90 > 90%, CD73 > 90%, CD31 < 2%, CD45 < 2%, and HLA-DR < 2%. The marker CD68 is used as macrophage identifier, CD163

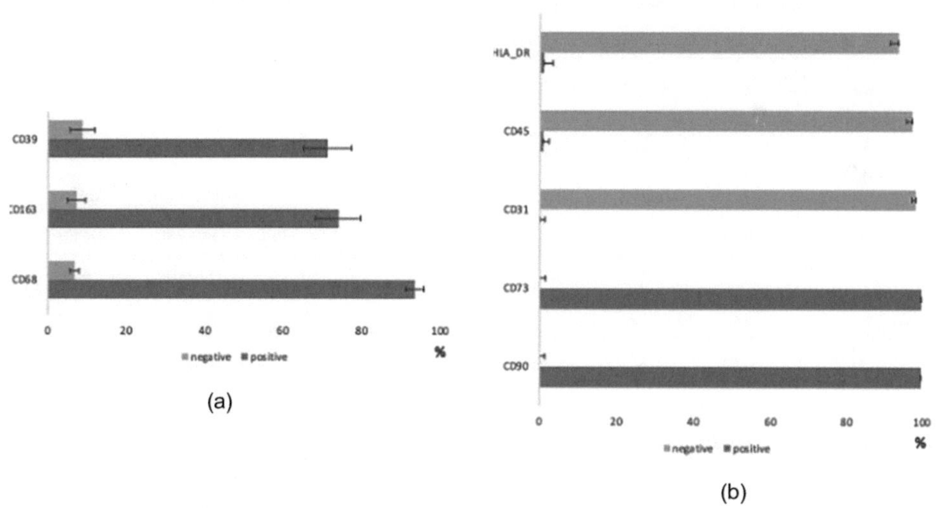

Figure 1.
Percentage of positive cells for each antibody tested.

Age	Sex	Complete blood (millions/µl)			Buffy volumen (ml)	PBMCs (millions)	Monocytes (millions)
		Luekocytes	Red blood cells	Platelets			
45	M	54.3	2.4	69	145	668.8	65.65

Table 3.
Data and yields obtained from monocyte donors.

is mostly present on M2 type macrophages, and CD39 is expressed on macrophages/ monocytes in co-culture with MSC.

3.2 Cytokine characterization of the conditioned medium

The results of quantification by both ELISA and Multiplex are detailed in **Table 4**.

To obtain the pattern (**Figure 2**), those molecules that could be quantified because they were within the detection limits of the method used in each case were studied, and statistically significant differences were sought with respect to the values taken as

CK		**MIP1-α**	**IL-2**	**IL-6**	**TIMP-1**	**IL-8**	**IL-10**
Monocytes	AVG	9.5643367	<8.2	<10.5	35777,0862	235.261369	3.06669833
	SD	3.57120227	NP	NP	48725,6036	106.055289	1.34471437
Control MSC	AVG	<2.3	<8.2	542.274142	>167500	9.30152294	0.7036029
Co-culture MSC/ monocytes	AVG	6.91058459	<8.2	1153.79376	>167500	330.463609	2.58248776
	SD	4.45252385	NP	309.210936	NP	175.835377	0.53200545
CK		IL-12 P70	IL-1 RA	RANTES	GMCSF	Leptina	HGF
Monocytes	AVG	2.08256741	476452.06	1.23761123	<17.3	49.1307951	1748.29405
	SD	0.04597914	826813.74	0.27823114	NP	1.9033707	412.510477
Control MSC	AVG	0.70858087	<41.4	4.7416231	<17.3	<16.3	145.833433
Co-culture MSC/ monocytes	AVG	1.68319311	223795.156	5.39009081	<17.3	64.0308586	1630.39368
	SD	0.30017417	208257.922	2.88381852	NP	49.0665959	456.036333
CK		MMP-3	MCP1	BNGF	EGF	Adiponectin	TNF-α
Monocytes	AVG	<16.5	2658.5485	<7.4	1.93098264	189.27	<5.9
	SD	NP	1672.2932	NP	NP	111.942	NP
Control MSC	AVG	137.490877	573.667894	<7.4	<1.2	<89.3	<5.9
Co-culture MSC/ monocytes	AVG	450.455082	3275.6692	0.48007142	<2.2	173.5	<5.9
	SD	322.742794	1773.64942	0.19422895	NP	23.5476	NP
CK		MMP-1	TRAIL	FGF-2	PDGF-BB	VEGF-A	IGF-1
Monocytes	AVG	32.0031459	1.63442629	<3	71.944999	15.6273891	760.452793
	SD	41.5287708	1.1152979	NP	37.2898086	26.2176915	1099.67079
Control MSC	AVG	560.670891	<1.2	<3	83.1238663	1838.37263	717.691553
Co-culture MSC/ monocytes	AVG	1364.87744	3.18084865	<3	127.82401	239.64569	1903.91696
	SD	335.663559	2.00494659	NP	80.6592455	399,735639	881.409224
CK		BMP-6	IL-1β	IL-4	TGF-β1	TGF-β3	VEGF-C
Monocytes	AVG	<156	<1	<10	<31.2	<31.2	<62.5
	SD	NP	NP	NP	NP	NP	NP
Control MSC	AVG	<156	<1	<10	<31.2	<31.2	<62.5

DOI: http://dx.doi.org/10.5772/intechopen.104486

CK		MIP1-α	IL-2	IL-6	TIMP-1	IL-8	IL-10
Co-culture MSC/ monocytes	AVG	<156	<1	<10	<31.2	<31.2	<62.5
	SD	NP	NP	NP	NP	NP	NP

Values are shown in picograms per milliliter. > and < indicate that the value is above or below the detection limits (respectively). SD: standard deviation; NP: not applicable.

Table 4.
Mean values of the molecules studied.

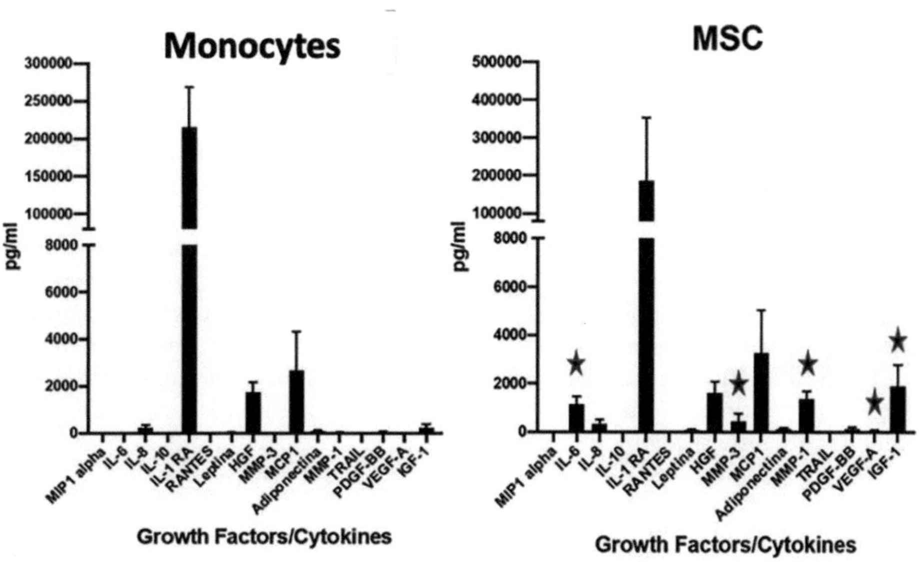

Figure 2.
Cytokine expression patterns. Shown are 16 mean values ± standard deviation of molecular characterization (MIP-1α, IL-6, IL-8, IL-10, IL-1Ra, RANTES, Leptine, HGF, MMP-3, MCP-1, Adiponectine, MMP-1, TRAIL, PDGF-BB, VEGF-A, IGF-1). (A) Secretome of monocytes; (B) secretome of co-culture. Stars mark values where there is a statistically significant difference ($p < 0.05$).

control (conditioned medium of M2-like monocytes/macrophages). Those values that were significantly different in all the samples studied (five samples in triplicate) were considered to form a specific and reproducible pattern of monocyte secretome modification by co-culture with MSCs. To test whether the pattern obtained was specific to the cell type co-cultured with monocytes, the expression of the same secretome molecules was obtained under the same conditions, but co-culturing monocytes with the following cell types: osteocytes, chondrocytes, tenocytes, synoviocytes, myocytes, lymphatic vascular cells, and Schwann cells, was compared. From this comparison, eight different and characteristic secretomes could be specifically differentiated, quantifying a minimum of seven molecules: IL-6, leptin, HGF, MMP-1, MMP-3, adiponectin, and VEGF-A (data not shown).

3.3 In vitro anti-inflammatory potency test

THP-1 cells after 48 h of stimulation with PMA gain adherence to the culture plastic and take on a macrophage-like morphology. After 24 h of culture, the number

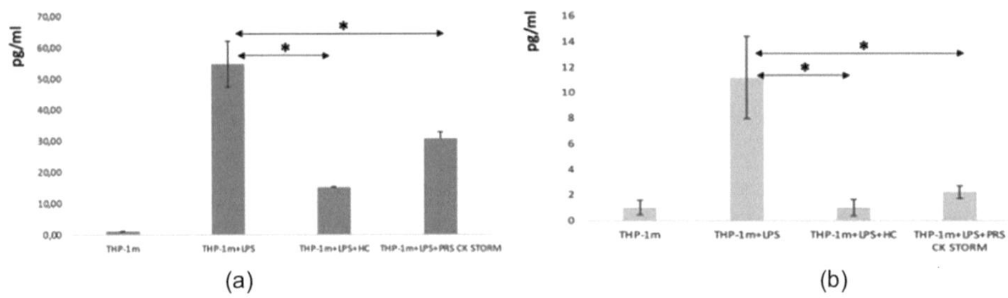

Figure 3.
Graph a shows the results in terms of IL-1β release in the in vitro model of inflammation. Graph b shows the results in terms of TNF-α release in the in vitro model of inflammation. For these four studies, the results were analyzed in three independent experiments with two replicates of the analytical technique, for each type of experiment. Error bars indicate the standard deviation between samples. Asterisks (*) mark values where there is a statistically significant difference ($p < 0.05$).

of cells in suspension decreases and the number of cells adherent to the plastic increases, a symptom of correct differentiation. After 48 h, about 90% of the cells are adherent to the plastic and are used in the experimental model.

The in vitro inflammation model used in this research is based on stimulating the proinflammatory action of macrophages when exposed to LPS. **Figure 3** shows that the addition of our conditioned medium to the culture of THP-1 cells transformed into macrophages can reverse the effect of LPS on these macrophages, and a statistically significant difference can be observed, in addition to the difference observed with the use of soluble hydrocortisone. It can also be observed that the cells are sensitive to LPS stimuli at the concentration used.

4. Biosafety and in vivo efficacy tests

The results of the Irwin test are analyzed to evaluate at a general level the effect produced by the conditioned medium administered intravenously to mice, in which the cytokine storm model had been previously induced, by retrorbital injection of LPS. A slight decrease in temperature was observed in all LPS-injected groups. LPS administration was observed to induce changes in reflexes and behavior, such as hunching, piloerection, and tremors, which were increased throughout the 3-day trial. In contrast, exploratory activity, reaction to contact, and aggressiveness were slightly decreased. In all LPS-treated mice (with or without drug administration), mild diarrhea occurred, which in untreated or gold standard-treated mice resulted in more severe dehydration than in mice treated with the conditioned medium. In general, there is a trend that the symptoms caused by LPS administration are dissipated by treatment with our conditioned medium (PRS® CK STORM), demonstrating a beneficial effect on the mice without counterproductive effects (**Table 5**).

The variations observed have as reference value the parameters 1 day before generating the model. Mice 11, 18, and 22 died before completing the 3 days of the experiment.

From the blood obtained after euthanasia and exsanguination of the animals, biochemical tests were performed to determine the biochemical profiles, which are shown in **Table 6**. The Mann–Whitney test was used for statistical analysis, considering statistical significance as$^*p < 0.05$ in the case of statistically significant minor differences with respect to the baseline value and $^{**}p < 0.05$ in the case of statistically significant major differences with respect to the baseline value.

	STANDARD		SLIGHT DECREASE
	SLIGHT INCREASE		MODERATE INCREASE
	MODERATE DECREASE		HIGH DECREASE
	HIGH INCREASE		DEAD

		CONTROL GROUP (BASAL)					CONTROL GROUP (PBS)					LPS					LPS+Gold Standard					LPS+PRS® CK STORM				
	Temperature (ºC)	1	2	3	4	5	6	7	8	9	10	11	12	13	14	15	16	17	18	19	20	21	22	23	24	25
	Pre treatment													36.1												
	20 minutes after LPS	-	-	-	-	-	-	-	-	-	-								36.2							
	48 hours after LPS*	36.9	37.6	36.7	37.5	37.5	36.8	37.3	36.7	37.0	37.2			34.9					34.2					35.2		
		CONTROL GROUP (BASAL)					CONTROL GROUP (PBS)					LPS					LPS+Gold Standard					LPS+PRS® CK STORM				
	Behaviour	1	2	3	4	5	6	7	8	9	10	11	12	13	14	15	16	17	18	19	20	21	22	23	24	25
Without manipulation	Stopping																									
	Agitation																									
	Piloerection																									
	Alarm																									
	Locomotor activity																									
	Stereotypes																									
	Front leg movement																									
	Head movement																									
	Chewing																									
	Ptosis																									
	Narrowed eyes																									
	Respiratory rate																									
	Vocalizations																									
	Tremors																									
	Seizures																									
Coordination	Ataxia																									
	View																									
	Grip strength																									
	Righting reflex																									
	Exploratory activity																									
	Straub sign																									
	Tail reflex																									
Manipulation	Contact reaction																									
	Abdominal tone																									
	Extremities tone																									
	Aggressiveness																									
	Pupil size																									
	Grip strength																									
	Tail suspension																									
	Acoustic reflex																									
	Lachrymation																									
	Salivation																									
	Urination																									
	Defecation																									
	Diarrhea																									
	Vocalizations																									
Others	Body temperature																									
	Dehydration																									
	Comments	All good, no increase in nervousness or aggressiveness in the bucket or in contact compared to yesterday																								

Table 5.
Results of the Irwin test; it includes the parameters studied on the third day after generating the model. Mice 11, 18 and 22 died before completing the 3 days of the experiment [52].

Figure 4 shows in percentage the relative variations observed at the posttreatment times (24, 48, and 72 h), with respect to the baseline values observed in each group.

The experimental model employed uses LPS as a causative agent of acute lung damage, causing a cytokine storm in the organism of mice like that produced by COVID-19 disease. We focused on the quantification of a small amount of these cytokines (TNF-α, IL-1β, IL-6, and IL-10) to evaluate the effect of the PRS® CK STORM. **Figure 5** shows the evolution of these cytokines detected in the sera of the mice on each of the days that the treatment lasted.

Histopathological analysis of the samples obtained from various organs of the mice obtained after the corresponding necropsies showed patchy lung thickening of the interstitium in a large part of the sample in the LPS treatment, while in the group treated with PRS® CK STORM it was observed that there was no lung damage, as in the control groups. Slight affectations were observed in liver and spleen, which the drug was also able to reverse. As for the heart and kidney, no pathological findings were detected. **Figure 6** shows examples of the lung sections studied in the different groups of the experiment.

4.1 Mechanism of action study

In order to approach the mechanism of action of our conditioned medium PRS® CK STORM, qPCR study is performed in order to analyze the effects of our complex biological drug under investigation on all common molecules involved in TLR2, TLR3, TLR4, TLR7, TLR8, TLR9, NOD1, NOD2, and NLRP3 pathways, TLR9, NOD1, NOD2, and NLRP3, and it is found that the conditioned medium downregulates the hyperactivity of these pathways, immunoregulating the key proteins involved in these pathways, being very remarkable the decrease of expression observed in TRAF6, caspase-1, RIPK1, IKKB, NF-kβ, MyD88, and NLRP3 proteins.

Following the method described in the corresponding section of this chapter, the total mRNA of the proteins named in the previous paragraph common to various pattern recognition pathways was extracted from THP-1m cells in culture used as control, from

			Albumin	Alanineaminotranferase	Amylase	Totalbilirrubin	Creatinine	Glucose	Total protein	Globulin
Control	0 h	Average	41	23	945	4	34	20.3	52	12
		SD	3.5	3.1	20.1	24	9.3	2.2	2.5	4.2
	24 h	Average	40	25.8**	946	7.0**	21.0*	22	55	15.0**
		SD	4.1	3.7	23.7	2.8	10.9	2.6	2.9	5
	48 h	Average	48.0**	30.0**	912	10.0**	22.0*	21.4	58.0**	10.0*
		SD	3	2.7	17.3	2.1	8	1.9	2.1	3.6
	72 h	Average	40	25	960	5.0**	27.0*	23.7**	52	13
		SD	3.1	2.8	17.8	2.1	8.2	1.9	2.2	3.7
PBS	0 h	Average	43	30	869	5	18	20.9	54	11
		SD	2.2	3.4	29.5	2.4	3.4	2.5	1.5	1.5
	24 h	Average	40	28	866	6.0**	24.0**	22.8	54	14.0**
		SD	2.5	4.3	34.9	2.8	4	3	1.8	1.8
	48 h	Average	38.0*	28	910	6.0**	25.0**	19.9	52	14.0**
		SD	1.9	2.9	25.5	2.1	2.9	2.2	1.3	1.3
	72 h	Average	39	29	924.5	5	25.0*	25.6**	51	12
		SD	1.9	3	162.3	2.1	3	2.2	1.3	1.3
LPS	0 h	Average	33	33	758	5	20	13.1	50	17
		SD	1.6	3	71.1	3	1.4	2	1.3	2.4
	24 h	Average	29.0*	26.0*	615.0*	5	15.9	10.5*	51	22.0**
		SD	1.9	3.5	83.9	3.6	1.7	2.3	1.5	2.8
	48 h	Average	31	18.0*	768	6.0**	14.5	9.0*	49	17
		SD	1.4	2.6	61.3	2.6	1.2	1.7	1.1	2.1
	72 h	Average	31	84.0**	740	7.0**	16	8.8*	48	18

			Albumin	Alanineaminotranferase	Amylase	Totalbilirrubin	Creatinine	Glucose	Total protein	Globulin
		SD	1.4	2.6	63	2.7	1.3	1.8	1.1	2.1
LPS + gold standard	0 h	Average	32	21	727	5	16.1	11	50	18
		SD	1.8	8.9	199.5	2.2	3.1	1.1	1.5	0.5
	24 h	Average	30	26.4**	854.0**	4.0*	15.4	9.5*	48	18
		SD	2.2	10.5	135.6	2.6	3.7	1.2	1.8	0.6
	48 h	Average	33	15.0*	1118.0**	5	14.6	9.0*	51	18
		SD	1.6	7.7	171.8	1.9	2.7	0.9	1.3	0.4
	72 h	Average	29	36.0**	1278.0**	9.0**	16	8.6*	48	19
		SD	1.6	7.9	176.6	2	2.8	0.9	1.3	0.4
PRS® CK STORM	0 h	Average	29.5	48.5	787.5	8	23.5	15	49	19
		SD	1	17.8	118	2.6	7.1	3.3	2	3.1
	24 h	Average	31.5	32.5*	681.5*	5.0*	29.0**	12.9*	49	17.0 *
		SD	1.2	2.1	139.2	3.1	8.3	3.9	2.4	3.6
	48 h	Average	30	36.0*	823	9.0**	21.5	15.3	50.5	19
		SD	0.9	1.5	101.6	2.3	6.1	2.8	1.7	2.6
	72 h	Average	38.5**	48	783.5	6.5*	21.5	14.5	52.5	18
		SD	0.9	1.6	104.5	2.3	6.3	2.9	1.8	2.7

*The Mann–Whitney test was used for the statistics, considering the statistical significance as *$p < 0.05$ in the case of minor statistically significant differences with respect to the basal value and **$p < 0.05$ in the case of major statistically significant differences with respect to the basal value.*

Table 6.
Results of the biochemical analysis of the mice, including the parameters studied pretreatment and posttreatment at 24, 48, and 72 h.

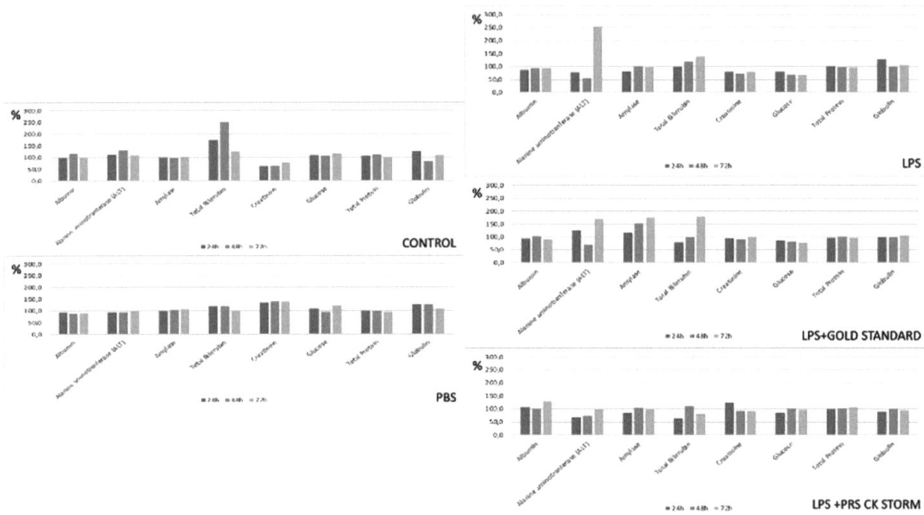

Figure 4.
Concentration values of the different metabolites that make up the biochemical profile of the mice, expressed as a percentage with respect to the baseline value observed in each group.

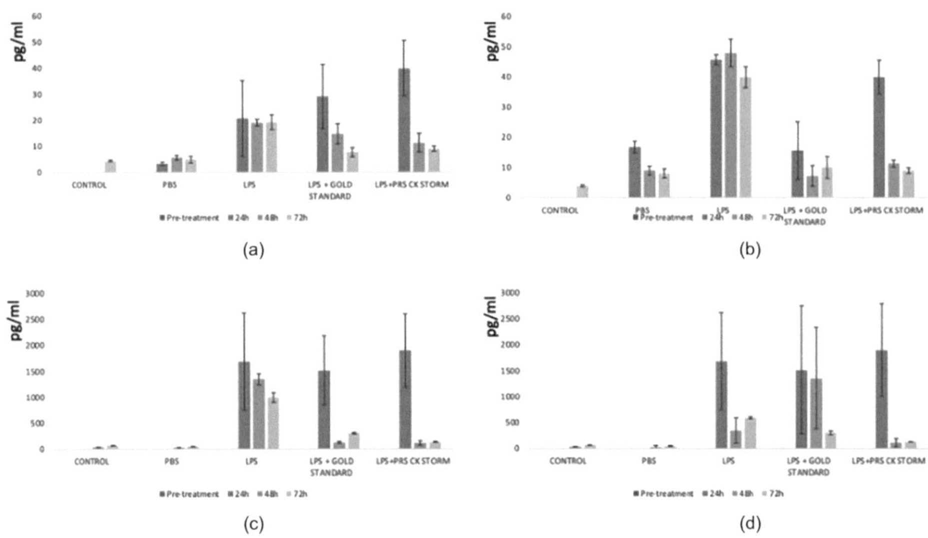

Figure 5.
Serum values of the different cytokines after 24, 48, and 72 h from the administration of the first treatment expressed in pg/ml as the mean of the values of the mice in each of the experimental groups. a: TNF-α. b: IL-1β. c: IL-6. d: IL-10. Values not shown are lower than the detection limit of the assay (2.8 pg/ml).

those stimulated with LPS, and from those stimulated with LPS and treated with PRS® CK STORM, obtaining by qPCR the relative expression of the genes at the mRNA level normalized against GAPDH, the results of which are shown in **Figure 7**.

Similarly and under the same experimental methodology, the total mRNA of the pattern recognition proteins linked to main infectious processes TLR-2, TLR-3, TLR-4, and TLR-7 was extracted from THP-1m cells in culture used as control, from those stimulated with LPS and from those stimulated with LPS and treated with PRS® CK

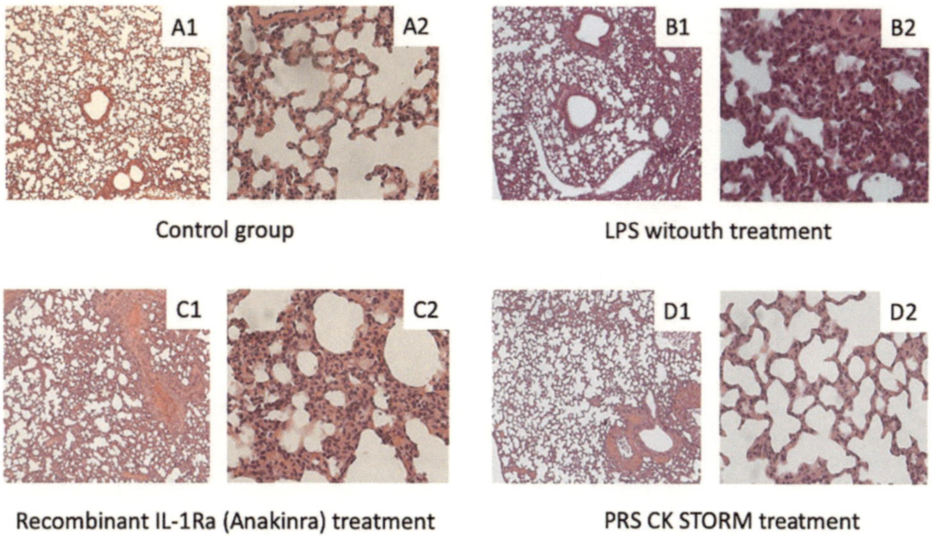

Figure 6.
Pulmonary anatomopathological study in the different groups of mice in the experiment. It can be seen that the group treated with PRS® CK STORM (our conditioned medium), the appearance of the lung is very similar to that shown by the untreated control group. All sections have been stained with hematoxylin eosin. ×1 images are taken in MO at 4× magnification and ×2 images are taken in MO at 40× magnification.

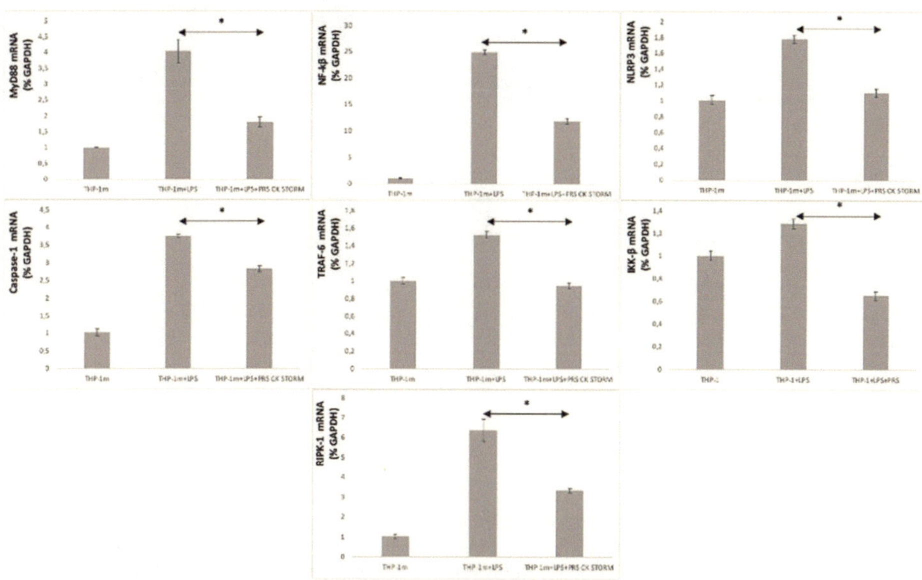

Figure 7.
*qPCR study showing the results of relative gene expression of proteins common to various pathways related to pattern recognition in infections, at the mRNA level normalized against GAPDH, considering statistical significance as *p < 0.05.*

STORM, obtaining by qPCR the relative expression of the genes at mRNA level normalized against GAPDH, whose results are shown in **Figure 8**.

In order to estimate the possible action of our PRS® CK STORM conditioned medium through the purinergic system (**Figure 9**), the results of the ATP/ADP ratio

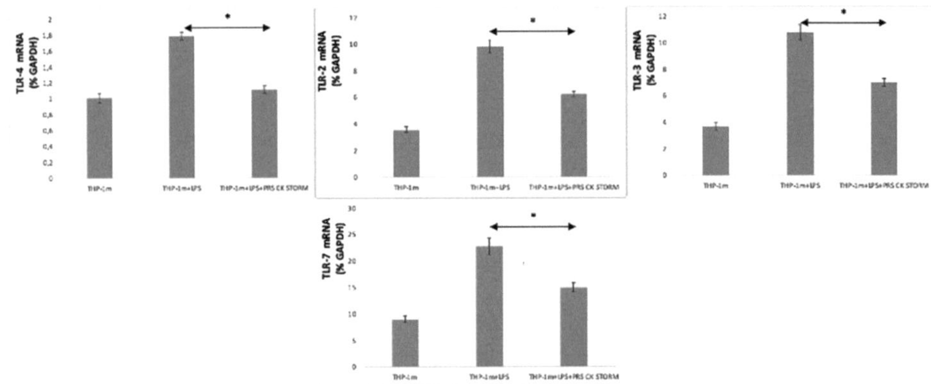

Figure 8.
*qPCR study showing the results of relative gene expression of major pattern recognition receptor proteins in infections, at the mRNA level normalized against GAPDH, considering statistical significance as *$p < 0.05$.*

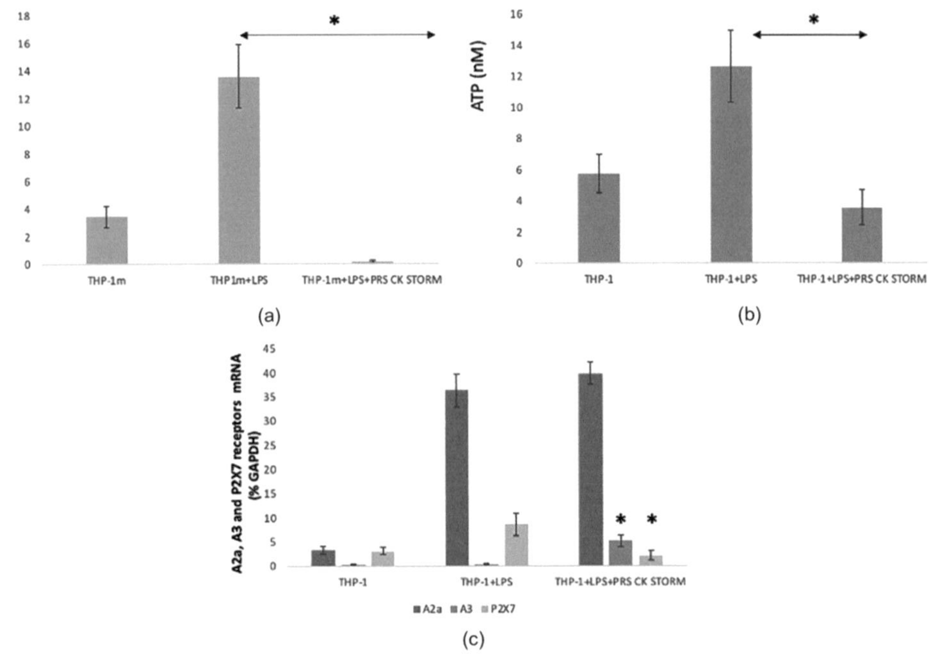

Figure 9.
*(a) ATP/ADP ratio analysis in three groups of cell cultures (THP-1m, THP-1m + LPS, and THP-1m + LPS + PRS® CK STORM). Asterisks (*) mark values where there is a statistically significant difference ($p < 0.05$). (b) Quantitative analysis of extracellular ATP in three groups of cell cultures (THP-1m, THP-1m + LPS, and THP-1m + LPS + PRS® CK STORM). Asterisks (*) mark values where there is a statistically significant difference ($p < 0.05$). (c) qPCR study showing the results of the relative expression of purinergic A2a, A3, and P2X7 receptor genes at the mRNA level normalized against GAPDH, considering statistical significance as *$p < 0.05$.*

obtained were analyzed comparatively among three groups with three replicates for each group, being the first group formed by THP-1m cells placed in culture alone, THP-1m cells stimulated with LPS at the doses described in Section 2, and the same THP-1m cells stimulated with LPS but in culture in the PRS® CK STORM conditioned medium (**Figure 9a**). The extracellular ATP contained in the same three groups of cultures with three replicates for each group was quantified (**Figure 9b**). Finally, following the method described in the corresponding section of this chapter, the total mRNA of purinergic receptors A2a, A3, and P2X7 was extracted from THP-1m cells in culture used as control, from those stimulated with LPS, and from those stimulated with LPS and treated with PRS® CK STORM, obtaining by qPCR the relative expression of the genes at the mRNA level normalized against GAPDH (**Figure 9c**).

5. Discussion

In the complex composition of this conditioned medium (PRS® CK STORM), all the growth factors, cytokines, and chemokines that are naturally produced by M2-type macrophages and MSCs, associated with innate immunity, are present, respecting the natural pleiotropic relationships, with an immunomodulatory cytokine profile from which a potent anti-inflammatory action is expected. In addition, according to the results obtained in test studies, the mechanism of action on TLR-7 receptors may possibly include some antiviral activity [32]. In fact, several experiments have shown that the secretomes of both cell types possess immunomodulatory properties. For example, direct injection of the supernatant of cultured mesenchymal stem cells (MSCs), which contains a variety of growth factors, prostaglandins and cytokines, can be applied to the treatment of kidney injury [45].

The theoretical advantage of using the complete conditioned medium versus some of its purified components lies in the synergistic mechanism between its different components [48], the result of subjecting the cell populations to a culture that, in vitro, attempts to emulate the immunomodulatory and regenerative microenvironment that occurs in vivo in diseased tissue. However, the main handicap of biological drugs with complex natural compositions will always be the variability between different batches and the practical impossibility of achieving a complete characterization, both qualitatively and quantitatively, as well as functionally [48]. Despite the earlier-mentioned, our group has been able to prove the existence of a stable cytokine pattern or fingerprint, which depends directly on the type of co-culture established and the conditions of the same and not on the donor of origin.

The anti-inflammatory capacity and potency shown by the PRS® CK STORM conditioned medium has been remarkable, showing statistically significant reductions in in vitro tests on TNF-α and IL-1β levels, these differences being very similar to those obtained with hydrocortisone. This anti-inflammatory immunomodulatory capacity, we have also been able to verify in the in vivo model, generated in mice. The mice treated with our PRS® CK STORM conditioned medium have shown normal behavior and the rest of the parameters analyzed in the Irwin test have been very similar to the control groups, where the cytokine storm had not been provoked. In fact, the comparative results in the in vivo test between the gold standard treatment used (Anakinra) and the PRS® CK STORM were clearly more favorable to the latter. It should be noted that in the group treated with PRS® CK STORM practically no pulmonary lesions were observed, while in the group treated with Anakinra inflammatory and fibrotic infiltrates were observed in a minimum of 30% of the surface of the sections. This coincides with the observation made at the time of sacrifice of the animals under study where both the control animals, in which the cytokine storm had not been provoked, and those treated with PRS® CK STORM, took more than 2 min to die in the CO_2 chamber, while the animals treated with Anakinra died in 40 s and those not treated in about 30 s, and these times can be directly related to the anatomopathological state observed at the pulmonary level.

The results obtained throughout this experiment suggest that the drug PRS® CK STORM is safe for intravenous administration, since no significant adverse effects have been observed in the different parameters analyzed, those found being mild. In addition, the drug significantly attenuates the detrimental effects caused by the cytokine storm associated with LPS administration. Most of the proteins and metabolites analyzed follow the same trend, regardless of the experimental group observed. With respect to albumin, the main protein present in blood, there is a decrease in albumin in all groups to which LPS was administered, which fits with that described by Ballmer et al. [53], with hypoalbuminemia occurring when the organism undergoes sepsis due to infection. Related to this, there is a decrease in total protein: a decrease in albumin will cause a decrease in total protein since the former is at very high concentrations. A decrease in blood glucose is also observed in all mice treated with LPS and LPS + gold standard. However, those treated with PRS® CK STORM managed to normalize the levels of glycemia, albuminemia, and total proteinemia 3 days after receiving the first dose. This fact was directly related to the clinical improvement observed in these animals subjected to the experimental treatment, given that they felt better than the rest of the mice, ate better, physical activity was normalized, and diarrhea was corrected. On the other hand, a significant increase in alanine amino transferase (ALT), total bilirubin, and amylase was observed during the experiment, which could be related to reactive hepatitis [54]. This increase was observed in all groups stimulated with LPS and was maintained in those treated with gold standard. However, in those treated with PRS® CK STORM, the figures normalized 72 h after the first treatment.

In all the groups where LPS was administered, there was a rapid increase in all the cytokines analyzed. From this, we can conclude that stimulation with bacterial lipopolysaccharide (LPS) is able to induce the expected inflammatory response, being more pronounced in more acute phase, the day after treatment administration and decreasing over time. In the control group and the vehicle, the presence of proinflammatory cytokines is not observed, which confirms that LPS is the cause of this response. However, in the group treated with LPS + gold standard, a lower increase of IL-1β is observed with respect to the rest of the groups where LPS was administered. This fact can be explained by the previous administration of the gold standard, recombinant IL-1Ra. However, despite the lower increase in this proinflammatory cytokine, the decrease in IL-1β finally achieved by the gold standard is even lower than that obtained with PRS® CK STORM treatment. In general terms, our PRS® CK STORM conditioned medium achieves greater control of all the cytokines analyzed in the experiment.

The results of the mechanism of action study show that PRS® CK STORM is able to immunomodulate in an anti-inflammatory way the expression of TLR-like pattern recognition pathways especially associated with infectious processes, such as TLR-2, TLR-3, TLR-4, and TLR-7. From the same study, it can be deduced that this effect is not only localized to these receptors but also acts at the level of various proteins common to these and other pathways, such as TRAF6, RIPK1, and IKKB, with the decrease in expression observed in the proteins NF-kβ, MyD88, caspase-1, and NLRP3 being very significant.

Many published studies have demonstrated the importance of the purinergic system in the inflammation associated with the cytokine storm caused by moderate/severe infection, including COVID-19, and have shown that using various purinergic system receptors as a therapeutic target can limit the negative effects of the cytokine storm [21, 22, 55–57]. Extracellular ATP at high concentrations becomes a true alarmin [58], a potent proinflammatory signal capable of overexpressing and stimulating P2X-type purinergic receptors, especially P2X7R, located on various immune cells (neutrophils, eosinophils, monocytes, macrophages, mast cells, and lymphocytes) [59]. Extracellular adenosine triphosphate (eATP) is a well-characterized DAMP that modulates function and plasticity [60, 61]. This nucleotide can be released by stressed, injured, and dying cells or in response to TLR activation, reaching high concentrations in the extracellular medium [62].

In contrast, it has been shown that the balance between ATP and adenosine concentration is crucial in immune homeostasis. CD39 and CD73 are two ectonucleotidases that cooperate in the generation of extracellular adenosine by ATP hydrolysis, thus tipping the balance toward immunosuppressive microenvironments. Extracellular adenosine through A2A receptor stimulation has the ability to prevent activation and proliferation of both macrophages and T cells, thereby dramatically decreasing cytokine production [63].

From the results observed both in the cytometric characterization of M2 macrophages used in the co-culture to produce our conditioned medium, where a strong expression of CD39 and CD73 is observed, and from the study of the ATP/ADP ratio, where a clear increase of ADP is observed, we can deduce that one of the mechanisms of action of PRS® CK STORM is probably linked to the process of dephosphorylation of extracellular ATP, which is degraded by ectonucleotidases to adenosine, and the latter interacts with adenosine receptors, type A2A and A3, producing an immunomodulatory anti-inflammatory effect on the cytokine storm. This theory is supported by two further pieces of evidence; firstly by the statistically significant decrease observed in LPS-stimulated THP-1 cells treated with our PRS® CK STORM conditioned medium with respect to the levels observed in untreated LPS-stimulated THP-1 cells; and secondly by the combination of the observed reduction in the relative expression of P2X7 receptor mRNA with the observed increase in the relative expression of A3 and A2a receptor mRNA in LPS-stimulated THP-1 cells treated with PRS® CK STORM relative to untreated THP-1 cells.

In all assays used in the in vitro and in vivo models, both employed LPS as an inducer of inflammation. Although the immunomodulatory mechanisms induced by bacterial antigens with respect to viral antigens in immune cells are different at the mechanistic level, the inflammatory response at the level of innate immunity ends up sharing numerous points in common [64, 65]. Therefore, the effect that PRS® CK STORM has on these models of inflammation with LPS can be exportable to what can happen, at the level of immune response, during a viral infection. In fact, it has been shown that monocytes and macrophages stimulated with LPS and ATP increase the release of IL-1β [66, 67].

All these results observed in the study on the possible mechanism of action of our complete conditioned medium (PRS® CK STORM) demonstrate that the immunomodulatory anti-inflammatory effect observed is a direct consequence of the action of various molecules contained in this medium, acting in a synergistic and pleiotropic manner on various therapeutic targets associated with different proinflammatory pathways, managing to downregulate the activation of the inflammasome, the inflammatory activation of the purinergic system, the activation of various pathways of pattern recognition associated with infections, etc., thus avoiding the possible feedback effects that therapeutic approaches based on the inhibition of a single receptor or a single inflammatory pathway may have.

6. Conclusions

The co-culture of M2 macrophages with MSCs allows the simple production of a complete conditioned medium (PRS® CK STORM), which has a clear anti-inflammatory profile. In line with this characterization, it has been demonstrated that PRS® CK STORM is able to stop macrophage overactivation in an in vitro inflammation model, through several mechanisms, including the expression pattern of TLR's and the purinergic system. Likewise, the action capacity of this drug has also been demonstrated in vivo, by improving the symptomatology and tissue damage induced in mice in another model of inflammation. Therefore, PRS® CK STORM is proposed as an effective and safe treatment to treat cytokine storms associated with moderate/severe infectious processes of any etiology, including that associated with COVID-19.

Author details

Juan Pedro Lapuente
SCO of the Living Cells Molecular and Cellular Biology Research Laboratory of the Fuenlabrada Universitary Hospital, Madrid, Spain

*Address all correspondence to: jplapuente@yahoo.es

References

[1] Weiss U. Nature insight: Inflammation. Nature. 2002;**420**:845

[2] Medzhitov R. Origin and physiological roles of inflammation. Nature. 2008;**454**:428-435

[3] Risdall RJ, McKenna RW, Nesbit ME, Krivit W, Balfour HH Jr, Simmons RL, et al. Virus-associated hemophagocytic syndrome: A benign histiocytic proliferation distinct from malignant histiocytosis. Cancer. 1979;**44**:993-1002

[4] Guo XJ, Thomas PG. New fronts emerge in the influenza cytokine storm. Seminars in Immunopathology. 2017;**39** (5):541-550. DOI: 10.1007/s00281-017-0636-y

[5] Tseng Y-T, Sheng W-H, Lin B-H, Lin C-W, Wang J-T, Chen Y-C, et al. Causes, clinical symptoms, and outcomes of infectious diseases associated with hemophagocytic lymphohistiocytosis in Taiwanese adults. Journal of Microbiology, Immunology, and Infection. 2011;**44**(3):191-197

[6] Cascio A, Pernice LM, Barberi G, Delfino D, Biondo C, Beninati C, et al. Secondary hemophagocytic lymphohistiocytosis in zoonoses. A systematic review. European Review for Medical and Pharmacological Sciences. 2012;**16**:1324-1337

[7] Singh ZN, Rakheja A, Yadav TP, Shome J. Infection-associated haemophagocytosis: The tropical spectrum. Clinical and Laboratory Haematology. 2005;**27**(5):312-315

[8] Teijaro JR. Cytokine storms in infection diseases. Seminars in Immunopathology. 2017;**39**:501-503

[9] Himanshu K, Taro K, Shizuo A. Pathogen recognition by the innate immune system. International Reviews of Immunology. 2011;**30**(1):16-34

[10] Taro K, Shizuo A. Toll-like receptors and their crosstalk with other innate receptors in infection and immunity. Immunity. 2011;**34**(5):637-650

[11] Komal D, Manoj B, Gourango B, Atul B, Sangita B. TLRs/NLRs: Shaping the landscape of host immunity. International Reviews of Immunology. 2011;**37**(1):3-19

[12] Martinon F, Burns K, Tschopp J. The inflammasome: A molecular platform triggering activation of inflammatory caspases and processing of proIL-beta. Molecular Cell. 2002;**10**(2):417-426

[13] Root-Bernstein R. Synergistic activation of toll-like and NOD receptors by complementary antigens as facilitators of autoimmune disease: Review, model and novel predictions. International Journal of Molecular Sciences. 2020;**21**:4645

[14] Hosseini AM, Majidi J, Baradaran B, Yousefi M. Toll-like receptors in the pathogenesis of autoimmune diseases. Advanced Pharmaceutical Bulletin. 2015; **5**:605-614

[15] Dabbagh K, Lewis DB. Toll-like receptors and T-helper-1/T-helper-2 responses. Current Opinion in Infectious Diseases. 2003;**16**:199-204

[16] Tukhvatulin AI, Gitlin II, Shcheblyakov DV, Artemicheva NM, Burdelya LG, Shmarov MM, et al. Combined stimulation of toll-like receptor 5 and NOD1 strongly potentiates activity of NF-kβ, resulting in enhanced innate immune reactions and resistance to *Salmonella enterica*

Serovar Typhimurium infection. Infection and Immunity. 2013;**81**: 3855-3864

[17] Moreira LO, Zamboni DS. NOD1 and NOD2 signaling in infection and inflammation. Frontiers in Immunology. 2012;**3**:328

[18] Olbei M, Hautefort I, Modos D, Treveil A, Poletti M, Gul L, et al. SARS-CoV-2 causes a different cytokine response compared to other cytokine storm-causing respiratory viruses in severely ill patients. Frontiers in Immunology. 2021;**1**(12):629193

[19] Root-Bernstein R. Innate receptor activation patterns involving TLR and NLR synergisms in COVID-19, ALI/ARDS and sepsis cytokine storms: A review and model making novel predictions and therapeutic suggestions. International Journal of Molecular Sciences. 2021;**22**:2108

[20] Simões JLB, Bagatini MD. Purinergic signaling of ATP in COVID-19 associated Guillain–Barré syndrome. Journal of Neuroimmune Pharmacology. 2021; **16**(1):48-58

[21] Franciosi MLM, Lima MDM, Schetinger MRC, Cardoso AM. Possible role of purinergic signaling in COVID-19. Molecular and Cellular Biochemistry. 2021;**476**(8):2891-2898

[22] Simões JLB, de Araújo JB, Bagatini MD. Anti-inflammatory therapy by cholinergic and purinergic modulation in multiple sclerosis associated with SARS-CoV-2 infection. Molecular Neurobiology. 2021;**58**(10): 5090-5111

[23] Abbracchio MP, Burnstock G. Purinoceptors: Are there families of P2X and P2Y purinoceptors? Pharmacology & Therapeutics. 1994;**64**:445-475

[24] Di Virgilio F, Dal Ben D, Sarti AC, Giuliani AL, Falzoni S. The P2X7 receptor in infection and inflammation. Immunity. 2017;**47**:15-31

[25] Di Virgilio F, Schmalzing G, Markwardt F. The elusive P2X7 macropore. Trends in Cell Biology. 2018; **28**:392-404

[26] Illes P, Rubini P, Ulrich H, Zhao Y, Tang Y. Regulation of microglial functions by purinergic mechanisms in the healthy and diseased CNS. Cell. 2020;**9**:1108

[27] Csoka B, Németh ZH, Törö G, Idzko M, Zech A, Koscsó B, et al. Extracellular ATP protects against sepsis through macrophage P2X7 purinergic receptors by enhancing intracellular bacterial killing. The FASEB Journal. 2015;**29**: 3626-3637

[28] Wu X, Ren J, Chen G, Wu L, Song X, Li G, et al. Systemic blockade of P2X7 receptor protects against sepsis-induced intestinal barrier disruption. Scientific Reports. 2017;7:4364

[29] Di Virgilio F, Tang Y, Sarti AC, Rossato M. A rationale for targeting the P2X7 receptor in coronavirus disease 19. British Journal of Pharmacology. 2020; **177**(21):4990-4994

[30] Ren W, Rubini P, Tang Y, Engel T, Illes P. Inherent P2X7 receptors regulate macrophage functions during inflammatory diseases. International Journal of Molecular Sciences. 2021;**23**(1):232

[31] Carty F, Mahon BP, English K. The influence of macrophages on mesenchymal stromal cell therapy: Passive or aggressive agents? Clinical and Experimental Immunology. 2017;**188**(1): 1-11

[32] Asami T, Ishii M, Fujii H, Namkoong H, Tasaka S, Matsushita K, et al. Modulation of murine macrophage TLR7/8-mediated cytokine expression by mesenchymal stem cell-conditioned medium. Mediators of Inflammation. 2013;**2013**:264260

[33] Vallés G, Bensiamar F, Crespo L, Arruebo M, Vilaboa N, Saldaña L. Topographical cues regulate the crosstalk between MSCs and macrophages. Biomaterials. 2015; **37**:124-133

[34] English K. Mechanisms of mesenchymal stromal cell immunomodulation. Immunology and Cell Biology. 2013;**91**(1):19-26

[35] Deng Y, Zhang Y, Ye L, Zhang T, Cheng J, Chen G, et al. Umbilical cord-derived mesenchymal stem cells instruct monocytes towards an IL10-producing phenotype by secreting IL6 and HGF. Scientific Reports. 2016;**6**: 37566

[36] Melief SM, Geutskens SB, Fibbe WE, Roelofs H. Multipotent stromal cells skew monocytes towards an anti-inflammatory interleukin-10-producing phenotype by production of interleukin-6. Haematologica. 2013;**98**:888-895

[37] Saldaña L, Bensiamar F, Vallés G, Mancebo FJ, García-Rey E, Vilaboa N. Immunoregulatory potential of mesenchymal stem cells following activation by macrophage-derived soluble factors. Stem Cell Research & Therapy. 2019;**10**(1):58

[38] Luz-Crawford P, Djouad F, Toupet K, Bony C, Franquesa M, Hoogduijn MJ, et al. Mesenchymal stem cell-derived interleukin 1 receptor antagonist promotes macrophage polarization and inhibits B cell differentiation. Stem Cells. 2016;**34**:483-492

[39] Ge W, Jiang J, Arp J, Liu W, Garcia B, Wang H. Regulatory T-cell generation and kidney allograft tolerance induced by mesenchymal stem cells associated with indoleamine 2,3-dioxygenase expression. Transplantation. 2010;**90**: 1312-1320

[40] Jiang XX, Zhang Y, Liu B, Zhang SX, Wu Y, Yu XD, et al. Human mesenchymal stem cells inhibit differentiation and function of monocyte-derived dendritic cells. Blood. 2005;**105**:4120-4126

[41] Ramasamy R, Fazekasova H, Lam EW, Soeiro I, Lombardi G, Dazzi F. Mesenchymal stem cells inhibit dendritic cell differentiation and function by preventing entry into the cell cycle. Transplantation. 2007;**83**:71-76

[42] Jackson MV, Morrison TJ, Doherty DF, McAuley DF, Matthay MA, Kissenpfennig A, et al. Mitochondrial transfer via tunneling nanotubes is an important mechanism by which mesenchymal stem cells enhance macrophage phagocytosis in the in vitro and in vivo models of ARDS. Stem Cells. 2016;**34**:2210-2223

[43] Gonçalves FDC, Luk F, Korevaar SS, Bouzid R, Paz AH, López-Iglesias C, et al. Membrane particles generated from mesenchymal stromal cells modulate immune responses by selective targeting of pro-inflammatory monocytes. Scientific Reports. 2017;**7**:12100

[44] Burnstock G, Boeynaems JM. Purinergic signalling and immune cells. Purinergic Signal. 2014;**10**:529-564

[45] Iseri K, Iyoda M, Ohtaki H, Matsumoto K, Wada Y, Suzuki T, et al. Therapeutic effects and mechanism of conditioned media from human mesenchymal stem cells on anti-GBM glomerulonephritis in WKY rats.

American Journal of Physiology. Renal Physiology. 2016;**310**(11):F1182-F1191

[46] Shen Y, Song J, Wang Y, Chen Z, Zhang L, Yu J, et al. M2 macrophages promote pulmonary endothelial cells regeneration in sepsis-induced acute lung injury. Annals of Translational Medicine. 2019;7(7):142

[47] Dahbour S, Jamali F, Alhattab D, Al-Radaideh A, Ababneh O, Al-Ryalat N, et al. Mesenchymal stem cells and conditioned media in the treatment of multiple sclerosis patients: Clinical, ophthalmological and radiological assessments of safety and efficacy. CNS Neuroscience & Therapeutics. 2017;**23**(11):866-874

[48] Laggner M, Gugerell A, Bachmann C, Hofbauer H, Vorstandlechner V, Seibold M, et al. Reproducibility of GMP-compliant production of therapeutic stressed peripheral blood mononuclear cell-derived secretomes, a novel class of biological medicinal products. Stem Cell Research & Therapy. 2020;**11**(1):9

[49] Lapuente JP, Dos-Anjos S, Blázquez-Martínez A. Intra-articular infiltration of adipose-derived stromal vascular fraction cells slows the clinical progression of moderate-severe knee osteoarthritis: Hypothesis on the regulatory role of intra-articular adipose tissue. Journal of Orthopaedic Surgery and Research. 2020;**15**(1):137

[50] Park EK, Jung HS, Yang HI, Yoo MC, Kim C, Kim KS. Optimized THP-1 differentiation is required for the detection of responses to weak stimuli. Inflammation Research. 2007;**56**(1):45-50

[51] Stephens RS, Johnston L, Servinsky L, Kim BS, Damarla M. The tyrosine kinase inhibitor imatinib prevents lung injury and death after intravenous LPS in mice. Physiological Reports. 2015;**3**(11):e12589

[52] Mathiasen JR, Moser VC. The Irwin test and functional observational battery (FOB) for assessing the effects of compounds on behavior, physiology, and safety pharmacology in rodents. Current Protocols in Pharmacology. 2018;**83**(1):e43

[53] Ballmer PE, Ochsenbein AF, Schütz-Hofmann S. Transcapillary escape rate of albumin positively correlates with plasma albumin concentration in acute but not in chronic inflammatory disease. Metabolism. 1994;**43**(6):697-705

[54] Neurath MF. COVID-19: Biologic and immunosuppressive therapy in gastroenterology and hepatology. Nature Reviews. Gastroenterology & Hepatology. 2021;**18**(10):705-715

[55] Leão Batista Simões J, Fornari Basso H, Cristine Kosvoski G, Gavioli J, Marafon F, Elias Assmann C, et al. Targeting purinergic receptors to suppress the cytokine storm induced by SARS-CoV-2 infection in pulmonary tissue. International Immunopharmacology. 2021;**100**(108150)

[56] Dos Anjos F, Simões JLB, Assmann CE, Carvalho FB, Bagatini MD. Potential therapeutic role of purinergic receptors in cardiovascular disease mediated by SARS-CoV-2. Journal of Immunology Research. 2020;**1** 8632048

[57] Hasan D, Shono A, van Kalken CK, van der Spek PJ, Krenning EP, Kotani T. A novel definition and treatment of hyperinflammation in COVID-19 based on purinergic signalling. Purinergic Signal. 2022;**18**(1):13-59

[58] Lai CC, Shih TP, Ko WC, Tang HJ, Hsueh PR. Severe acute respiratory syndrome coronavirus 2 (SARS-CoV-2) and coronavirus disease-2019 (COVID-19): The epidemic and the challenges. International Journal of Antimicrobial Agents. 2020;**55**(3):105924

[59] Liu P, Chen W, Chen JP. Viral metagenomics revealed Sendai virus and coronavirus infection of Malayan pangolins (*Manis javanica*). Viruses. 2019;**11**(11):979

[60] Barberà-Cremades M, Baroja-Mazo A, Pelegrín P. Purinergic signaling during macrophage differentiation results in M2 alternative activated macrophages. Journal of Leukocyte Biology. 2016;**99**(2):289-299

[61] Savio LEB, Coutinho-Silva R. Immunomodulatory effects of P2X7 receptor in intracellular parasite infections. Current Opinion in Pharmacology. 2019;**47**:53-58

[62] Cohen HB, Briggs KT, Marino JP, Ravid K, Robson SC, Mosser DM. TLR stimulation initiates a CD39-based autoregulatory mechanism that limits macrophage inflammatory responses. Blood. 2013;**122**(11):1935-1945

[63] Bono MR, Fernández D, Flores-Santibáñez F, et al. CD73 and CD39 ectonucleotidases in T cell differentiation: Beyond immunosuppression. FEBS Letters. 2015;**589**:3454-3460

[64] Finlay BB, McFadden G. Anti-immunology: Evasion of the host immune system by bacterial and viral pathogens. Cell. 2006;**124**(4):767-782

[65] Goldstein ME, Scull MA. Modeling innate antiviral immunity in physiological context. Journal of Molecular Biology. 2021;**1**:167374

[66] Ferrari D, Chiozzi P, Falzoni S, Dal Susino M, Melchiorri L, Baricordi OR, et al. Extracellular ATP triggers IL-1 beta release by activating the purinergic P2Z receptor of human macrophages. Journal of Immunology. 1997;**159**(3):1451-1458

[67] Perregaux DG, McNiff P, Laliberte R, Conklyn M, Gabel CA. ATP acts as an agonist to promote stimulus-induced secretion of IL-1 beta and IL-18 in human blood. Journal of Immunology. 2000;**165**(8):4615-4623

Chapter 8

The Potential of the Purinergic System as a Therapeutic Target of Natural Compounds in Cutaneous Melanoma

Gilnei Bruno da Silva, Daiane Manica, Marcelo Moreno and Margarete Dulce Bagatini

Abstract

Cutaneous melanoma is an aggressive and difficult-to-treat disease that has rapidly grown worldwide. The pharmacotherapy available in so many cases results in low response and undesirable side effects, which impair the life quality of those affected. Several studies have been shown that the purinergic system is involved in cancer context, such as in cutaneous melanoma. With technological advances, several bioactive compounds from nature are studied and presented as promising adjuvant therapies against cancer, as phenolic compounds and related action by purinergic system modulations. Thus, phenolic compounds such as rosmarinic acid, resveratrol, tannic acid, as well as vitamin D may be promising substances in a therapeutic perspective to treat cutaneous melanoma *via* purinergic system pathway. More research needs to be done to open up new horizons in the treatment of melanoma by the purinergic signaling.

Keywords: purinergic signaling, skin cancer, adjuvant therapies, therapeutic target

1. Introduction

Cutaneous melanoma (CM) is a disease that arises in transition of dermis and epidermis, where the melanocytes are localized. The melanogenesis process starts due to DNA damage secondary to a UV exposition, which can be chronic or acute intermittent exposure. Furthermore, other risk factors are associated with melanoma as frequency of sunbathing, ultraviolet A exposure, low skin phototypes, atypical nevus syndrome, skin sunburn events mostly during childhood and adolescence, a large number of skin moles (congenital or not), familiar or personal history of CM or skin cancer not melanoma. The DNA damage alters the proliferation and cell cycle, culmining in dysregulated apoptosis mechanisms. The CM is characterized by the high invasiveness, a high metastatic capacity, causing a short survival period and high mortality rates due to pharmacological resistance [1, 2].

For the systemic treatment of the patient with high-risk disease to metastasis, pharmacotherapy is used with drugs that can manifest collateral effects, in addition to presenting inefficient mechanisms to guarantee the survival of patients [3]. In this sense, several literatures have indicated biochemical therapy as a promising adjuvant in the CM management [4], even so it is urgent researchers to develop new options for a rise in patients' survival [1]. Therefore, many research science teams have been engaged to discover melanoma treatment and an interesting alternative to this would be to use natural substances, such as phenolic compounds that can have anticancer effects [5].

An efficient and rapid diagnosis is a priority among the medical and scientific community [6]. Furthermore, correctly and effectively pharmacological therapy promotes better prognosis and better quality of life for the melanoma cutaneous illness. Taking into consideration, the aim of this chapter was to provide an overview of potential modulations of the purinergic system in the treatment of cutaneous melanoma. Thus, initially, the cutaneous melanoma will be characterized based on epidemiology and therapeutics, as well as the purinergic system with its details, and finally, show that some natural compounds have potential in modulating purinergic signaling in melanoma.

2. Cutaneous melanoma: Epidemiology and current therapeutic treatments

Cutaneous melanoma is a disease that has a wide spectrum in relation to the prognosis of the patient who develops this disease. The diagnosis can be made from when the neoplasm is still restricted to the epidermis (*in situ*) in the disseminated form, where the malignant disease can affect other organs [7]. Data from the latest Global Cancer Statistics survey estimated for the year 2020 that cutaneous melanoma represents 1.7% of all malignant neoplasms, corresponding to 324,635 new cases worldwide. The number of deaths due to melanoma was 57,043. Comparing these numbers with the previous survey (2018), there was a considerable increase in the number of new cases, which was 232,100, as well as the number of deaths recorded (55,550 deaths). Both incidence and mortality rates differ according to each region of the planet, in addition to regions within the same country. Oceania has the highest ratio of number of cases; in Australia for the year 2020, it was expected 1 case of melanoma for every 104 male inhabitants and 1 case for every 185 females [8].

Also, North America and some European countries have high incidence rates of this disease. In the United States alone, 99,780 new cases and 7650 deaths from melanoma are expected in 2022 [9]. In the Scandinavian region, the incidence ranges from 15 to 18 cases/100,000 inhabitants; Northern European countries vary from 12 to 28 cases/100,000 inhabitants [10]. In Asia, Africa, and South America, the number of cases is lower, although there are regional differences as is the case in Brazil, where the South and Southeast regions have the highest number of cases when compared with other areas of the country [11–14].

Early detection of cutaneous melanoma is a fundamental factor in reducing mortality. Individuals diagnosed in the early stages have a 98% survival rate, while those diagnosed in advanced stages have a significantly decreased survival rate—between 63.8 and 15%. Patients diagnosed with stage III and IV melanoma have survival rates (5 years) of 70% and 30%, respectively. For most human malignancies, the use of chemotherapy for systemic treatment changed the natural history of the disease, and in melanoma the reality, until recently, was different with response rates comparable to the use of placebo [15]. Since the introduction of chemotherapy for adjuvant treatment of malignant neoplasms, numerous therapeutic regimens have been tried in patients with metastatic melanoma [16, 17].

The most widely used treatment for patients with disseminated disease was dacarbazine, but only about 15–20% had some degree of response and 2% were still alive after 5 years of follow-up: a response rate comparable to the placebo-treated group of patients in the early clinical trials [15]. The use of high-dose interleukin-2 (IL-2) was the first treatment that changed the natural history of a small portion of patients with stage IV melanoma, but resulted in severe side effects that often affected survival [18, 19]. In the following years, molecules with direct action on pathways were responsible for controlling cell growth and division emerged, which were called "targeted therapy," such as BRAF and MEK inhibitors [19, 20]. However, only part of the melanoma patients can benefit from these inhibitors, because it is necessary that the genes involved in these pathways are mutated in order to get a response [21].

With the development of immunotherapy, other molecules called "immune system checkpoint inhibitors", such as pembrolizumab and nivolumab (programmed cell death protein-1 [PD-1] inhibitors), and ipilimumab (cytotoxic T lymphocyte antigen-4 [CTLA-4] inhibitor), have been routinely used as adjuvant and neoadjuvant treatments in melanoma patients with prognostic factors associated with poor survival [22, 23]. However, only 20% of patients show complete and lasting response with this type of therapy, and no biomarkers have been defined yet that can predict who will benefit from the use of these drugs [23–25].

3. Evidences that purinergic system plays a role in cutaneous melanoma

The purinergic system is a sophisticated cell-cell communication and ubiquitously expressed in the human body that orchestrates numerous cellular responses in the context of health and disease, displaying several biological processes. Most discussed extracellular signaling molecules include nucleotides such as adenosine triphosphate (ATP), adenosine diphosphate (ADP), adenosine monophosphate (AMP), and the nucleoside adenosine (Ado) [26–30]. The components belonging to this system are divided between receptors, according to the signaling molecule, as well as in enzymes. Thus, the P2 receptor is sensitized by adenine nucleotides, such as ATP, ADP, and AMP, being subdivided into P2X (1–7) and P2Y (1–12), while the P1 group is signaled Ado molecules and differentiated into A1, A2, A2B, A3 [31]. The levels of signaling molecules are controlled by enzymes known as ectonucleotidases, expressed on the surface of cells. They are nucleoside triphosphate diphosphohydrolase 1 (NTPDase-1/CD39), 5′-nucleotidase (5'-NT/CD73), and adenosine deaminase (ADA) enzymes, which metabolize ATP/ADP into AMP, AMP into adenosine (Ado), and finally this into inosine, respectively [32–35] (**Figure 1**).

In this context, it is well known that this system is involved in cancer dynamics and has a close relationship with the immune responses, such as in lung cancer [36, 37], leukemia [38], cutaneous melanoma [30, 39], pancreatic cancer [40], and gastric cancer [41]. Likewise, it has been reported that ATP is considered a pivotal molecule, which largely influences immune responses in peripheral and central tissues, can be released from the inflammatory and tumor cells *via* different mechanisms, such as exocytosis, plasma-membrane channels pannexin, or lysis, and getting accumulated in tumor microenvironment (TME) [42–46].

Thus, yet few studies have been performed to understand the involvement of purinergic signaling in melanoma, some robust evidence showed that this cell pathway

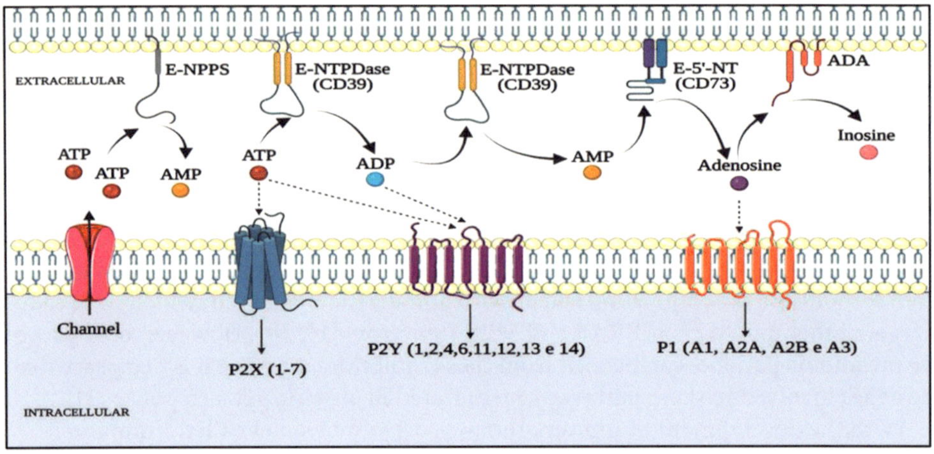

Figure 1.
Purinergic system components and functions. Adenosine triphosphate (ATP), key molecule of the purinergic system, can be released in an extra-cell environment and act as agonist on P2XR and P2YR. Furthermore, the ectonucleotidases presenting on cell surface are capable to hydrolyze the ATP into others nucleotides, such as adenosine monophosphate (AMP), adenosine diphosphate (ADP), and nucleosides, such as adenosine (ado). Between the purinergic receptors, only P1 (P1R) have Ado as agonist. The ectonucleotidase pyrophosphatase/ phosphodiesterase (E-NPPS) has potential to break ATP straight to AMP, whereas ectonucleoside triphosphate diphosphohydrolase (E-NTPDase-CD39) can break ATP to ADP or ADP to AMP. In this enzyme orchestra, only one ectoenzyme is capable of hydrolyzing AMP to Ado, the ecto-5′-nucleotidase (E-5'-NT-CD73). In the end of the purinergic cascade, the adenosine deaminase (ADA) converts Ado to inosine (Ino). Source: The authors (2022).

plays an important role in this disease. In this sense, a component-system widely reported is the P2X7 receptor, which is expressed by cancer cells, as in melanoma, and is mostly associated with tumor cell killing *via* ATP key molecule signaling modulation [47–49]. However, if hydrolyzed to ADP can present an immunosuppression effect [39], as well as Ado, a nucleoside product of ATP hydrolysis that mediates the protective response, such as immunosuppressive and anti-inflammatory effects on healthy tissues, has a pro-carcinogenic property on melanoma [50–53]. Data about the role of ATP in melanoma patients were found by Mânica et al. [50], who indicated that an increased inflammatory process by extracellular ATP leads to an immunosuppressive profile even after surgical removal, which, in turn, corroborated previous information.

Given the pleiotropic actions, the role of P2X7R depends on the nucleotide receptor-interactions, as well as concentration, being that these interactions can promote or inhibit melanoma. Taking account, recently it was reinforced that P2X7 is overexpressed in patients affected by metastatic malignant melanoma and that its expression closely correlates with reduced overall survival. This is because P2X7 stimulation is capable of miRNA-containing microvesicles and exosomes from melanoma cells [54]. Furthermore, it was hypothesized that the Warburg effect is possibly linked to P2X7 modulation by ATP in melanoma. Once activated by ATP, the PI3K-AKT pathway upregulates glycolytic cascade enzymes, which promotes lactate generation and acidification of the TME. Acidification of extracellular microenvironment alters immune response and supporting cancer [5].

Conversely, Hattori et al. [55] by treatment of B16 melanoma cells with oxidized ATP (oxATP) found significantly decreased cell proliferation at concentrations between 300 and 500 mM in low pH conditions. From this, they proposed that the P2X7R is a promising target for treatment of solid tumors. The same way, White et al. [56] after an experimental incubation of melanoma cells with P2X7-agonist 2′–3′-O-(4-benzoylbenzoyl) adenosine 5′-triphosphate found decrease in cell number. In the immunological scenario, P2X7 activity has been associated with tumor-infiltrating T cells (TILs), which leads to senescence and limits tumor suppression, in addition to affected cell cycling of effector T cells and resulting in generation of mitochondrial reactive oxygen species (ROS) and p38 MAPK-dependent upregulation of cyclin-dependent kinase inhibitor 1A [57].

Although the P2X are widely related, other receptors also have been involved in melanoma disease. The P2Y1 receptor was indicated as potential to reduce melanoma cell proliferation; however, P2Y2 usually appears to increase cell numbers [58]. Still, the P2Y12 seems to promote tumor metastasis by platelet activation in melanoma cells [59, 60]. Interestingly, one factor, which leads to skin cancer, UV-B irradiation, seems to have a relationship with purinergic signaling, and severe effects have been associated between irradiation type and reduced P2X1 and P2Y2 receptors, as well as to destruction of P2X7 receptors, with the possibility of contributing to malignant transformation of keratinocytes [61].

From the Ado-stimulated receptor perspective, the A2AR and A3AR seem to lead to melanoma cells' death *via* proliferation mechanisms. Deletion of A2ARs was capable of reverting immunosuppression in B16-melanoma-bearing mice, immune cells responses [62–64]. Koszałka et al. [65] also showed that A1R, A2AR, and A3R receptors play an important role in melanoma (B16 type cells) by modulating angiosupport and immunosuppression in mice. An interesting study performed with melanoma cells discovered that A2B receptor blockade can impair IL-8 production, whereas blocking A3 receptors, it is possible to further decrease VEGF secretion in melanoma cells treated with etoposide (VP-16) and doxorubicin. Thus, treatment of melanoma cells with the DNA-damaging drugs such as VP-16 and doxorubicin resulted in Ado receptors modulations and chemotherapeutic potential [66].

On the other side, ectonucleotidases that control the purinergic chain also are involved in neoplasias. The CD39 decreased activity mitigates ATP hydrolysis, leading to extracellular accumulation of this nucleotide. This was evidenced by Manica et al. [50] that showed that post-surgery CM patients present high ATP levels in microenvironment compared with the healthy controls, suggesting being the cause of poor prognosis [50]. Thus, the ectonucleotidases action seems to play an important role in cancer context as well as in other purinergic components.

Although the CD39 and CD73 dynamics are responsible for forming most Ado extracellular content, another enzyme involved is the E-NPP, which hydrolyzes AMP to Ado. Several studies have evidenced that these ectoenzymes are increased in cancer context [45, 67–69]. In melanoma, studies suggested high expression of the CD73 in patients [70–72]. Also, in the melanoma mouse model, a CD73 inhibitor improved T and B cell-mediated antitumor immunity and reduced tumor growth [73]. The hydrolysis CD73 capacity is known and involved in melanoma; however, recently a nonenzymatic action of this enzyme was related, playing a role in cell migration on extracellular matrix through focal adhesion kinase (FAK) [74].

4. Natural compounds with antitumor effect and their purinergic system relationship

Considering the need for new therapies and therapeutic targets for the treatment of cutaneous melanoma, studies with compounds that modulate the purinergic system and have antitumor effects have been carried out, as is the case with phenolic compounds and vitamin D [75, 76]. **Figure 2** represents some possible modulatory mechanisms of natural compounds on the purinergic system in the CM context.

Phenolic compounds are secondary metabolites present in plants, whose function is to participate in their development and protect them from pathogens and UV radiation [77]. More than 8000 compounds have been identified so far, and most of these compounds have some beneficial property to humans [78]. Their therapeutic actions are related to the structure, mainly to the phenolic rings, in which these compounds are classified by the number of rings and structural elements they have, forming four major groups: phenolic acids, stilbenes, lignans, and flavonoids [79]. The most cited group of polyphenols in the literature with therapeutic actions including antitumor effect is flavonoids, which are present in foods consumed daily such as fruits, vegetables, vegetables, red wine, coffee, and green tea [80–82].

Resveratrol (3,5,4-trihydroxy-trans-stilbene) is found in red wine and in the skin of dark grapes, a polyphenol with antitumor activity, considered a candidate for the treatment of cutaneous melanoma, which has been shown to modulate the expression and activity of CD73, ADA enzyme, and P2X7 and A2A receptors, which are closely related to tumor progression [83–86]. Tannic acid, a polyphenol, was shown to be able to induce cell death in several types of cancer cells, such as cutaneous melanoma, prostate cancer, glioblastoma [87, 88].

Thus, Bona et al. [89] tested the antitumor effect of this substance in rats with glioblastoma and the interference with ectonucleotidases, in which tannic acid was able to increase the hydrolysis of ATP and AMP nucleotides and decrease the

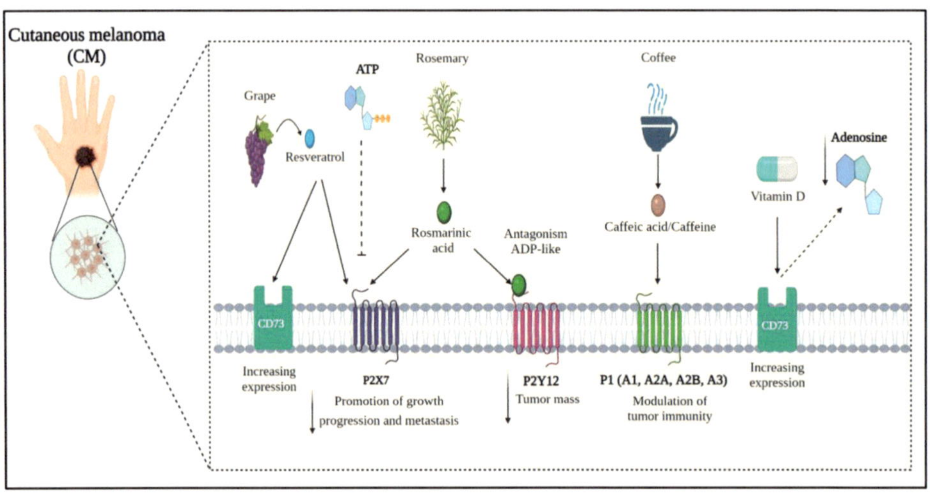

Figure 2.
Purinergic system modulation in cutaneous melanoma by means of natural compounds. The literature has been evidenced in the adjuvant therapeutic perspective that several compounds derived from nature can act against carcinogenesis and cutaneous melanoma. Resveratrol, from the grape, has the potential both to increase CD73 expression and modulate P2X7 receptors, which control the growth progression and metastasis. Rosmarinic acid, a phenolic acid from rosemary, can block the P2X7 receptor and inhibits the agonist mechanism by ATP, as well as have antagonism ADP-like on the P2Y12 receptor, leading to a decrease of tumor mass formation. Two important derivatives of coffee, the caffeic acid and caffeine, have been shown as interesting modulators of P1 receptors. These adenosinergic receptors are intimately related to tumor immunity, and these two molecules can act modulating the antitumor immunity. The vitamin D also showed a significant compound with purinergic system modulation, since it is capable of increasing CD73 expression and controlling the adenosine amount, which seems to play a great role in cutaneous melanoma. Source: The authors (2022).

hydrolysis of ADP in platelets of the animals treated compared with untreated. In the lymphocytes of the animals with the disease that received tannic acid, this polyphenol decreased the hydrolysis of ATP and ADP and the degradation of adenosine in relation to the group with the disease that did not receive the substance. When comparing the levels of ectonucleotidases in control mice, those with glioblastoma and those with glioblastoma treated with tannic acid, it was observed that the substance was able to maintain levels similar to those in mice without the disease. Bearing in mind that the purinergic system is able to modulate tumor progression, the aggressiveness of this type of cancer, and the results obtained in the study in question, tannic acid can be considered a promising agent for the treatment of cancer [89].

Regarding purinergic system modulation and the antitumor effect on cutaneous melanoma, Silva et al. [5] proposed the hypothesis that rosmarinic acid, a polyphenol with antitumor effect, would be able to modulate purinergic signaling and prevent tumor progression and metastasis by two-way means: by blocking the P2X7 receptor or by antagonizing the P2Y12 receptor. Interestingly, a paper that focused on Salvia *yunnanensis* extract, which contains rosmarinic acid in its composition, proved the inhibition ADP-induced of rabbit platelet aggregation by binding rosmarinic acid with P2Y12R [90].

Coffee (*Coffea arabica*) and green tea (*Camellia sinensis*) derivatives such as caffeine (1,3,7 trymetylxantine), caffeic acid (3,4-dihydroxycinnamic acid) and chlorogenic acid (3-O-caffeoylquinic acid) have shown promising effects in degenerative and cardiovascular diseases, in which they have been shown to modulate inflammation and purinergic signaling, mainly through the P1 family receptors that are closely related to tumor immunity, being strong candidates for the treatment of CM [91, 92]. To confirm the previous data, caffeic acid together with the antineoplastic dacarbazine decreased the viability of SK-Mel-28 metastatic cutaneous melanoma cells [93].

Quercetin, an abundant flavonoid in plants, also demonstrated antitumor activity in cell lines of bladder cancer, glioblastoma, and hepatocarcinoma and inhibited the activity and expression of ecto-5′- NT/CD73, leaving less Ado available in the TME, consequently preventing immunosuppression [94–96].

Apigenin (4′,5,7-trihydroxyflavone) is a flavonoid present in significant amounts in parsley, onion, celery, orange, chamomile, oregano, and basil that has shown a beneficial effect in diseases such as cancer, Alzheimer's, diabetes, and depression [97, 98]. Cutaneous melanoma cells (A375) were treated with these substance and, in addition to the decrease in cell viability, they had an increase in ATPase activity and a concomitant reduction in the ATP/ADP ratio related to the apoptotic process of cancer cells, demonstrating that the present substance has an effect antitumor in addition to acting in purinergic system [99].

The carcinogenesis can be initiated by the overproduction of reactive oxygen species (ROS), since the antioxidant defenses cannot neutralize these molecules. From this, the antioxidant compounds seem to be important against tumor formation, such as the phenolic acids, which can prevent DNA alterations and genome instability. The literature has shown that rosmarinic acid is an example of a powerful antioxidant for the protection of the DNA against UV and H_2O_2 [100].

Vitamin D, in turn, is a fat-soluble vitamin, in which its deficiency is closely related to carcinogenesis [101]. This vitamin is available in two forms: vitamin D2 (ergocalciferol) and vitamin D3 (cholecalciferol), where it can be obtained through the diet, but 90% of the daily needs are produced by the skin itself and later must be hydroxylated by the liver and kidneys, in order to obtain 1,25-dihydroxyvitamin D, the metabolically active form of vitamin D [102]. In metastatic melanoma cells, SK-Mel-28-treated with 1,25-dihydroxyvitamin D, there was decrease in cell viability, as well as the expression and activity of the CD73 enzyme and the levels of Ado, which has a suppressive function of tumor immunity and is essential for progression tumor, becoming a promising candidate for the adjuvant treatment of cutaneous melanoma [103].

5. Conclusion

As evidenced, cutaneous melanoma is a malignant neoplasm of great medical importance due to high rates of resistance to treatments and relapses, and for this reason, it is necessary to search for new and effective pharmacological therapies. In this context, the potential of some compounds in modulations of this pathway signaling, such as rosmarinic acid, resveratrol, tannic acid, as well as vitamin D, has been elucidated in this study. Of course, more research needs to be done to open up new horizons in the treatment of melanoma by the purinergic signaling, but the discovery of new ways to improve the anticancer pharmacological perspective has already begun.

Funding

MDB acknowledges grant support by Conselho Nacional de Desenvolvimento Científico e Tecnológico (CNPq, proj. No 310606/2021–7) and Fundação de Amparo à Pesquisa e Inovação do Estado de Santa Catarina (FAPESC, proj. No 2021TR1543). GBS, DM, MM, and MDB also thank Fundo de Apoio à Manutenção e ao Desenvolvimento da Educação Superior (FUMDES/UNIEDU) for the graduate scholarships.

Conflict of interest

The authors declare that there is no conflict of interest.

Author details

Gilnei Bruno da Silva, Daiane Manica, Marcelo Moreno
and Margarete Dulce Bagatini*
Graduate Program in Biomedical Sciences, Federal University of Fronteira Sul, Chapecó, SC, Brazil

*Address all correspondence to: margaretebagatini@yahoo.com.br

References

[1] Schadendorf D, et al. Melanoma. The Lancet. 2018;**392**(10151):971-984. DOI: 10.1016/S0140-6736(18)31559-9

[2] Paddock LE, et al. Skin self-examination and long-term melanoma survival. Melanoma Research. 2016;**26**(4):401-408. DOI: 10.1097/CMR.0000000000000255

[3] Siegel R, et al. Cancer statistics, 2013. CA: a Cancer Journal for Clinicians. 2013;**63**(1):11-30. DOI: 10.3322/caac.21166

[4] Wilson MA, Schuchter LM. Chemotherapy for melanoma. In: Kaufman HL, Mehnert JM, editors. Melanoma. Cancer Treatment and Research. Vol. v. 167. Cham: Springer International Publishing; 2016. pp. 209-229. DOI: 10.1007/978-3-319-22539-5_8

[5] da Silva GB, Yamauchi MA, Zanini D, Bagatini MD. Novel possibility for cutaneous melanoma treatment by means of rosmarinic acid action on purinergic signaling. Purinergic Signalling. 2022).;**18**:61-81. DOI: 10.1007/s11302-021-09821-7

[6] Mânica A, Bagatini MD. Melanoma cutâneo e sistema purinérgico. In: Cardoso AM, Manfredi LH, Maciel SFV d O, editors. Sinalização Purinérgica: Implicações Fisiopatológicas. Chapecó: UFFS; 2021. pp. 156-171

[7] Miller AJ, Mihm MC. Melanoma. The New England Journal of Medicine. 2006;**355**(1):51-65

[8] Australian Institute of Health and Welfare 2021. Cancer in Australia 2021. Cancer series nº. 133. Cat. nº. CAN 144. Canberra: AIHW

[9] Siegel RL, Miller KD, Fuchs HE, Jemal A. Cancer statistics, 2022. CA: a Cancer Journal for Clinicians. 2022;**72**(1):7-33

[10] Garbe C, Keim U, Eigentler TK, Amaral T, Katalinic A, Holleczek B, et al. Time trends in incidence and mortality of cutaneous melanoma in Germany. Journal of the European Academy of Dermatology and Venereology. 2019;**33**(7):1272-1280

[11] Sung H, Ferlay J, Siegel RL, Laversanne M, Soerjomataram I, Jemal A, et al. Global cancer statistics 2020: GLOBOCAN estimates of incidence and mortality worldwide for 36 cancers in 185 countries. CA: a Cancer Journal for Clinicians. 2021;**71**(3):209-249

[12] Olsen CM, Whiteman DC. Clinical epidemiology of melanoma. In: Balch CM, Atkins MB, Garbe C, Gershenwald JE, Halpern AC, Kirkwood JM, et al., editors. Cutaneous Melanoma. Cham: Springer International Publishing; 2020. pp. 425-449

[13] INCA - Instituto Nacional do Câncer (Brazil). Cancer of Skin Melanoma. Brasília, DF: Instituto Nacional do Câncer; 2021. Available from: https://www.inca.gov.br/tipos-de-cancer/cancer-de-pele-melanoma Accessed: March 2021

[14] Moreno M, Schmidt JC, Grosbelli L, Dassi M, Mierzwa RV. Análise de prevalência e mortalidade associada ao melanoma cutâneo em pacientes atendidos em centro de referência no Oeste do estado de Santa Catarina, Brasil, de 2002 a 2016. Rev Ciênc EM SAÚDE. 2020;**10**(4):109-116

[15] Yang AS, Chapman PB. The history and future of chemotherapy for melanoma. Hematology/

Oncology Clinics of North America. 2009;**23**(3):583-597

[16] Chabner BA, Roberts TG. Chemotherapy and the war on cancer. Nature Reviews. Cancer. 2005;**5**(1):65-72

[17] DeVita VT, Chu E. A history of cancer chemotherapy. Cancer Research. 2008;**68**(21):8643-8653

[18] Finn OJ. Cancer immunology. The New England Journal of Medicine. 2008;**358**(25):2704-2715

[19] Flaherty KT, Robert C, Hersey P, Nathan P, Garbe C, Milhem M, et al. Improved survival with MEK inhibition in BRAF-mutated melanoma. The New England Journal of Medicine. 2012;**367**(2):107-114

[20] Colombino M, Capone M, Lissia A, Cossu A, Rubino C, De Giorgi V, et al. BRAF/NRAS mutation frequencies among primary Tumors and metastases in patients with melanoma. Journal of Clinical Oncology. 2012;**30**(20):2522-2529

[21] Kwak EL, Clark JW, Chabner B. Targeted agents: The rules of combination: Table 1. Clinical Cancer Research. 2007;**13**(18):5232-5237

[22] Leach DR, Krummel MF, Allison JP. Enhancement of antitumor immunity by CTLA-4 blockade. Science. 1996;**271**(5256):1734-1736

[23] Nakamura Y. Biomarkers for immune checkpoint inhibitor-mediated tumor response and adverse events. Frontiers in Medicine. 2019;**6**:119

[24] Orloff M, Weight R, Valsecchi ME, Sato T. Immune check point inhibitors combination in melanoma: Worth the toxicity? Reviews on Recent Clinical Trials. 2016;**11**(2):81-86

[25] Havel JJ, Chowell D, Chan TA. The evolving landscape of biomarkers for checkpoint inhibitor immunotherapy. Nature Reviews. Cancer. 2019;**19**(3): 133-150

[26] Burnstock G, et al. Cellular distribution and function of P2 receptor subtypes in different systems. International Review of Cytology. 2004;**240**:31-304. DOI: 10.1016/S0074-7696(04)40002-3

[27] Atkinson B, et al. Ecto-nucleotidases of the CD39/NTPDase family modulate platelet activation and thrombus formation: Potential as therapeutic targets. Blood Cells, Molecules, and Diseases. 2006;**36**:217-222. DOI: 10.1016/j.bcmd.2005.12.025

[28] Burnstock G, et al. Purinergic Signaling, and the Nervous System. Berlin, Heidelberg: Springer; 2012. DOI: 10.1007/978-3-642-28863-0

[29] Visovatti SH, et al. Increased CD39 nucleotidase activity on microparticles from patients with idiopathic pulmonary arterial hypertension. PLoS One. 2012;**7**:408-429. DOI: 10.1371/journal.pone.0040829

[30] Bagatini MD, et al. The impact of purinergic system enzymes on noncommunicable, neurological, and degenerative diseases. Journal of Immunology Research. 2018;**2018**:1-21. DOI: 10.1155/2018/4892473

[31] Bartoli F, et al. Purinergic signaling and related biomarkers in depression. Brain Sciences. 2020;**10**(3):1-12

[32] Zimmermann H. Cellular function and molecular structure of ectonucleotidases. Purinergic Signalling. 2012;**8**:437-502. DOI: 10.1007/s11302-012-9309-4

[33] Yegutkin GG, et al. Nucleotide-and nucleoside-converting ectoenzymes:

Important modulators of purinergic signalling cascade. Biochimica et Biophysica Acta – Molecular Cell Research. 2008;**1783**:673-694. DOI: 10.1016/j.bbamcr.2008.01.024

[34] Schetinger MRC, et al. NTPDase and 5-nucleotidase activities in physiological and disease conditions: New perspectives for human health. BioFactors. 2007;**31**:77-98. DOI: 10.1002/biof.5520310205

[35] Zimmermann H. Ectonucleotidases in the nervous system. Novartis Foundation Symposium. 2006;**276**:113-128

[36] Zanini D, et al. Ectoenzymes and cholinesterase activity and biomarkers of oxidative stress in patients with lung cancer. Molecular and Cellular Biochemistry. 2013;**374**(1-2):137-148. DOI: 10.1007/s11010-012-1513-6

[37] Zanini D, et al. ADA activity is decreased in lymphocytes from patients with advanced stage of lung ca9ncer. Medical Oncology. 2019.;**36**:78. DOI: 10.1007/s12032-019-1301-1

[38] Ledderose C, et al. Cutting off the power: Inhibition of leukemia cell growth by pausing basal ATP release and P2X receptor signaling? Purinergic Signal. 2016;**12**(3):439-451. DOI: 10.1007/s11302-016-9510-y

[39] Mânica A, et al. The signaling effects of ATP on melanoma-like skin cancer. Cellular Signalling. 2019;**59**:122-130. DOI: 10.1016/j.cellsig.2019.03.021

[40] Hu L-P, et al. Targeting purinergic receptor P2Y2 prevents the growth of pancreatic ductal adenocarcinoma by inhibiting cancer cell glycolysis. Clinical Cancer Research. 2019;**25**(4):1318-1330. DOI: 10.1158/1078-0432.CCR-18-2297

[41] Hevia MJ, et al. Differential effects of purinergic Signaling in gastric cancer-derived cells through P2Y and P2X receptors. Frontiers in Pharmacology. 2019;**10**:612. DOI: 10.3389/fphar.2019.00612

[42] Burnstock G. Short- and long-term (trophic) purinergic signalling. Philosophical Transactions of the Royal Society B: Biological Sciences. 2016.;**371**:1700. DOI: 10.1098/rstb.2015.0422

[43] Stagg J, Smyth MJ. Extracellular adenosine triphosphate and adenosine in cancer. Oncogene. 2010;**29**(39):5346-5358. DOI: 10.1038/onc.2010.292

[44] Burnstock G. Blood cells: An historical account of the roles of purinergic signalling. Purinergic Signalling. 2015;**11**(4):411-434. DOI: 10.1007/s11302-015-9462-7

[45] Di Virgilio F, et al. Extracellular purines, purinergic receptors and tumor growth. Oncogene. 2017.;**36**(3):293-303. DOI: 10.1038/onc.2016.206

[46] Ghiringhelli F, et al. Activation of the NLRP3 inflammasome in dendritic cells induces IL-1β–dependent adaptive immunity against tumors. Nature Medicine. 2009;**15**(10):1170-1178. DOI: 10.1038/nm.2028

[47] Bian S, Sun X, Bai A, Zhang C, Li L, Enjyoji K, et al. P2X7 integrates PI3K/AKT and AMPK-PRAS40-mTOR signaling pathways to mediate tumor cell death. PLoS One. 2013;**8**:60184. DOI: 10.1371/journal.pone.0060184

[48] Di Virgilio F. Purines, purinergic receptors, and cancer. Cancer Research. 2012;**72**(21):5441-5447. DOI: 10.1158/0008-5472.CAN-12-1600

[49] Feng L, et al. Vascular CD39/ENTPD1 directly promotes tumor cell growth

by scavenging extracellular adenosine triphosphate. Neoplasia. 2011;**13**(3):206-216. DOI: 10.1593/neo.101332

[50] Mânica A, et al. High levels of extracellular ATP lead to chronic inflammatory response in melanoma patients. Journal of Cellular Biochemistry. 2018;**119**(5):3980-3988. DOI: 10.1002/jcb.26551

[51] Whintton B, et al. Vacuolar ATPase as a potential therapeutic target and mediator of treatment resistance in cancer. Cancer Medicine, 7, 8, p. 3800-3811, 2018. DOI: 10.1002/cam4.1594

[52] Di Virgilio F, et al. Extracellular ATP and P2 purinergic signalling in the tumour microenvironment. Nature Reviews Cancer. 2018;**18**(10):601-618. DOI: 10.1038/s41568-018-0037-0

[53] Pasquali S, et al. Systemic treatments for metastatic cutaneous melanoma. The Cochrane Database of Systematic Reviews. 2018;**2**(2):CD011123. DOI: 10.1002/14651858.CD011123.pub2

[54] Pegoraro A, et al. P2X7 promotes metastatic spreading and triggers release of miRNA-containing exosomes and microvesicles from melanoma cells. Cell Death & Disease. 2021;**12**:1088. DOI: 10.1038/s41419-021-04378-0

[55] Hattori F, et al. Feasibility study of B16 melanoma therapy using oxidized ATP to target purinergic receptor P2X7. European Journal of Pharmacology. 2012;**625**:20-26. DOI: 10.1016/j.ejphar.2012.09.001

[56] White N, et al. Human melanomas express functional P2X7 receptors. Cell and Tissue Research. 2005;**321**:411-418. DOI: 10.1007/s00441-005-1149-x

[57] Romagnani A, et al. P2X7 receptor activity limits accumulation of T cells within tumors. Cancer Research. 2020;**80**(18):3906-3919. DOI: 10.1158/0008-5472.CAN-19-3807

[58] White N, et al. P2Y purinergic receptors regulate the growth of human melanomas. Cancer Letters. 2005;**224**(1):81-91. DOI: 10.1016/j.canlet.2004.11.027

[59] Gebremeskel S, et al. The reversible P2Y12 inhibitor ticagrelor inhibits metastasis and improves survival in mouse models of cancer: P2Y12 inhibitor ticagrelor inhibits metastasis. International Journal of Cancer. 2015.;**136**(1):234-240. DOI: 10.1002/ijc.28947

[60] Kamiyama M, et al. ASK1 facilitates tumor metastasis through phosphorylation of an ADP receptor P2Y12 in platelets. Cell Death and Differentiation. 2017;**24**:2066-2076. DOI: 10.1038/cdd.2017.114

[61] Ruzsnavszky O, et al. UV-B induced alteration in purinergic receptors and signaling on HaCaT keratinocytes. Journal of Photochemistry and Photobiology B: Biology. 2011;**105**(1):113-118. DOI: 10.1016/j.jphotobiol.2011.07.009

[62] Merighi S, et al. Adenosine receptors as mediators of both cell proliferation and cell death of cultured human melanoma cells. Journal of Investigative Dermatology. 2002;**119**(4):923-933. DOI: 10.1046/j.1523-1747.2002.00111.x

[63] Cekic C, et al. Myeloid expression of adenosine A_{2A} receptor suppresses T and NK cell responses in the solid tumor microenvironment. Cancer Research. 2014;**74**(24):7250-7259. DOI: 10.1158/0008-5472.CAN-13-3583

[64] Fishman P, et al. A3 adenosine receptor as a target for cancer therapy.

Anti-Cancer Drugs. 2002;**13**(5):437-443. DOI: 10.1097/00001813-200206000-00001

[65] Koszałka P, et al. Specific activation of A3, A2A and A1 adenosine receptors in CD73-knockout mice affects B16F10 melanoma growth, neovascularization, angiogenesis and macrophage infiltration. PLoS One. 2016.;**11**(3):e0151420. DOI: 10.1371/journal.pone.0151420

[66] Merighi S, et al. A2B and A3 adenosine receptors modulate vascular endothelial growth factor and interleukin-8 expression in human melanoma cells treated with etoposide and doxorubicin. Neoplasia. 2009;**11**:1064-1073. DOI: 10.1593/neo.09768

[67] Aliagas E, et al. High expression of ecto-nucleotidases CD39 and CD73 in human endometrial tumors. Mediators of Inflammation. 2014;**2014**:509027. DOI: 10.1155/2014/509027

[68] Cappellari AR, et al. Characterization of ectonucleotidases in human medulloblastoma cell lines: Ecto-5'NT/CD73 in metastasis as potential prognostic factor. PLoS One. 2012;7:e47468. DOI: 10.1371/journal.pone.0047468

[69] Longhi MS, et al. Biological functions of ecto-enzymes in regulating extracellular adenosine levels in neoplastic and inflammatory disease states. Journal of Molecular Medicine (Berlin, Germany). 2013;**91**:165-172. DOI: 10.1007/s00109-012-0991-z

[70] Jiang T, et al. Comprehensive evaluation of NT5E/CD73 expression and its prognostic significance in distinct types of cancers. BMC Cancer. 2018;**18**(1):267. DOI: 10.1186/s12885-018-4073-7

[71] Sadej R, et al. Expression of ecto-5′-nucleotidase (eN, CD73) in cell lines from various stages of human melanoma. Melanoma Research. 2006;**16**(3):213-222. DOI: 10.1097/01.cmr.0000215030.69823.11

[72] I. Monteiro, et al. CD73 expression and clinical significance in human metastatic melanoma. Oncotarget, 9, 42, p. 26659-26669. 2018. DOI: 10.18632/oncotarget.25426.

[73] Forte G, et al. Inhibition of CD73 improves B cell-mediated anti-tumor immunity in a mouse model of melanoma. The Journal of Immunology. 2012;**189**(5):2226-2233. DOI: 10.4049/jimmunol.1200744

[74] Sadej R, et al. Dual, enzymatic and non-enzymatic, function of ecto-5′-nucleotidase (eN, CD73) in migration and invasion of A375 melanoma cells. Acta Biochimica Polonica. 2012;**59**(4):647-652

[75] Petric R, Braicu C, Raduly L, Dragos N, Dumitrascu D, Berindan-Negoe I, et al. Phytochemicals modulate carcinogenic signaling pathways in breast and hormone-related cancers. Onco Targets and Therapy. 2015;**8**:2053. DOI: 10.2147/OTT.S83597

[76] Niedzwiecki A, Roomi M, Kalinovsky T, Rath M. Anticancer efficacy of polyphenols and their combinations. Nutrients. 2016;**8**(9):552. DOI: 10.3390/nu8090552

[77] Kozikowski AP, Tückmantel W, Böttcher G, Romanczyk LJ. Studies in polyphenol chemistry and bioactivity. 4. [1] Synthesis of trimeric, tetrameric, pentameric, and higher oligomeric epicatechin-derived procyanidins having all-4β,8-interflavan connectivity and their inhibition of cancer cell growth through cell cycle arrest [1]. The Journal

of Organic Chemistry. 2003;**68**(5):1641-1658. DOI: 10.1021/jo020393f

[78] Santos-Buelga C, González-Paramás AM, Oludemi T, Ayuda-Durán B, González-Manzano S. Plant phenolics as functional food ingredients. Advances in Food and Nutrition Research. 2019;**90**:183-257. DOI: 10.1016/bs.afnr.2019.02.012

[79] Manach C, Scalbert A, Morand C, Rémésy C, Jiménez L. Polyphenols: Food sources and bioavailability. The American Journal of Clinical Nutrition. 2004;**79**(5):727-747. DOI: 10.1093/ajcn/79.5.727

[80] Lakenbrink C, Lapczynski S, Maiwald B, Engelhardt UH. Flavonoids and other polyphenols in consumer brews of tea and other caffeinated beverages. Journal of Agricultural and Food Chemistry. 2000;**48**(7):2848-2852. DOI: 10.1021/jf9908042

[81] Rodríguez-García C, Sánchez-Quesada C, Gaforio JJ. Dietary flavonoids as cancer chemopreventive agents: An updated review of human studies. Antioxidants. 2019;**8**(5):137. DOI: 10.3390/antiox8050137

[82] Kopustinskiene DM, Jakstas V, Savickas A, Bernatoniene J. Flavonoids as anticancer agents. Nutrients. 2020;**12**(2):457. DOI: 10.3390/nu12020457

[83] Bottari NB, Pillat MM, Schetinger MRC, Reichert KP, Machado V, Assmann CE, et al. Resveratrol-mediated reversal of changes in purinergic signaling and immune response induced by Toxoplasma gondii infection of neural progenitor cells. Purinergic Signalling. 2019;**15**(1):77-84. DOI: 10.1007/s11302-018-9634-3

[84] Fracasso M, Reichert K, Bottari NB, da Silva AD, Schetinger MRC, Monteiro SG, et al. Involvement of ectonucleotidases and purinergic receptor expression during acute Chagas disease in the cortex of mice treated with resveratrol and benznidazole. Purinergic Signalling. 2021;**17**(3):493-502. DOI: 10.1007/s11302-021-09803-9

[85] Marinheiro D, Ferreira B, Oskoei P, Oliveira H, Daniel-da-Silva A. Encapsulation and enhanced release of resveratrol from mesoporous silica nanoparticles for melanoma therapy. Materials. 2021;**14**(6):1382. DOI: 10.3390/ma14061382

[86] Sánchez-Melgar A, Muñoz-López S, Albasanz JL, Martín M. Antitumoral action of resveratrol through adenosinergic signaling in C6 glioma cells. Frontiers in Neuroscience. 2021;**15**:702817. DOI: 10.3389/fnins.2021.702817

[87] Bridgeman CJ, Nguyen T-U, Kishore V. Anticancer efficacy of tannic acid is dependent on the stiffness of the underlying matrix. Journal of Biomaterials Science, Polymer Edition. 2018;**29**(4):412-427. DOI: 10.1080/09205063.2017.1421349

[88] Nagesh PKB, Chowdhury P, Hatami E, Jain S, Dan N, Kashyap VK, et al. Tannic acid inhibits lipid metabolism and induce ROS in prostate cancer cells. Scientific Reports. 2020;**10**(1):980. DOI: 10.1038/s41598-020-57932-9

[89] Bona NP, Soares MSP, Pedra NS, Spohr L, da Silva dos Santos F, de Farias AS, et al. Tannic acid attenuates peripheral and brain changes in a preclinical rat model of glioblastoma by modulating oxidative stress and purinergic signaling. Neurochemical Research [Internet]. 2022;**47**:1541-1552. DOI: 10.1007/s11064-022-03547-7

[90] Li Y, et al. Screening for the antiplatelet aggregation quality markers of Salvia yunnanensis based on an integrated approach. Journal of Pharmaceutical and Biomedical Analysis. 2020;**188**:113383. DOI: 10.1016/j.jpba.2020.113383

[91] Stefanello N, Spanevello RM, Passamonti S, Porciúncula L, Bonan CD, Olabiyi AA, et al. Coffee, caffeine, chlorogenic acid, and the purinergic system. Food and Chemical Toxicology. 2019;**123**:298-313. DOI: 10.1016/j.fct.2018.10.005

[92] Castro MFV, Stefanello N, Assmann CE, Baldissarelli J, Bagatini MD, da Silva AD, et al. Modulatory effects of caffeic acid on purinergic and cholinergic systems and oxi-inflammatory parameters of streptozotocin-induced diabetic rats. Life Sciences. 2021;**277**:119421. DOI: 10.1016/j.lfs.2021.119421

[93] Pelinson LP, Assmann CE, Palma TV, da Cruz IBM, Pillat MM, Mânica A, et al. Antiproliferative and apoptotic effects of caffeic acid on SK-Mel-28 human melanoma cancer cells. Molecular Biology Reports. 2019;**46**(2):2085-2092. DOI: 10.1007/s11033-019-04658-1

[94] Braganhol E, Tamajusuku ASK, Bernardi A, Wink MR, Battastini AMO. Ecto-5′-nucleotidase/CD73 inhibition by quercetin in the human U138MG glioma cell line. Biochimica et Biophysica Acta (BBA) - General Subjects. 2007;**1770**(9):1352-1359. DOI: 10.1016/j.bbagen.2007.06.003

[95] Rockenbach L, Bavaresco L, Fernandes Farias P, Cappellari AR, Barrios CH, Bueno Morrone F, et al. Alterations in the extracellular catabolism of nucleotides are involved in the antiproliferative effect of quercetin in human bladder cancer T24 cells. Urologic Oncology: Seminars and Original Investigations. 2013;**31**(7):1204-1211. DOI: 10.1016/j.urolonc.2011.10.009

[96] Abruzzese V, Matera I, Martinelli F, Carmosino M, Koshal P, Milella L, et al. Effect of quercetin on ABCC6 transporter: Implication in HepG2 migration. International Journal of Molecular Sciences. 2021;**22**(8):3871. DOI: 10.3390/ijms22083871

[97] Hostetler GL, Ralston RA, Schwartz SJ. Flavones: Food sources, bioavailability, metabolism, and bioactivity. Advances in Nutrition. 2017;**8**(3):423-435. DOI: 10.3945/an.116.012948

[98] Salehi B, Venditti A, Sharifi-Rad M, Kręgiel D, Sharifi-Rad J, Durazzo A, et al. The therapeutic potential of apigenin. International Journal of Molecular Sciences. 2019;**20**(6):1305. DOI: 10.3390/ijms20061305

[99] Das S, Das J, Samadder A, Boujedaini N, Khuda-Bukhsh AR. Apigenin-induced apoptosis in A375 and A549 cells through selective action and dysfunction of mitochondria. Experimental Biology and Medicine (Maywood, N.J.). 2012;**237**(12):1433-1448. DOI: 10.1258/ebm.2012.012148

[100] Sevgi K, et al. Antioxidant and DNA damage protection potentials of selected phenolic acids. Food and Chemical Toxicology: An International Journal Published for the British Industrial Biological Research Association. 2015;**77**:12-21. DOI: 10.1016/j.fct.2014.12.006

[101] Holick MF. Vitamin D: Its role in cancer prevention and treatment. Progress in Biophysics and Molecular Biology. 2006;**92**(1):49-59. DOI: 10.1016/j.pbiomolbio.2006.02.014

[102] Lips P. Vitamin D physiology. Progress in Biophysics and Molecular Biology. 2006;**92**(1):4-8. DOI: 10.1016/j.pbiomolbio.2006.02.016

[103] Bagatini MD, Bertolin K, Bridi A, Pelinson LP, da Silva Rosa Bonadiman B, Pillat MM, et al. 1α, 25-Dihydroxyvitamin D3 alters ectonucleotidase expression and activity in human cutaneous melanoma cells. Journal of Cellular Biochemistry. 2019;**120**(6):9992-10000. DOI: 10.1002/jcb.28281